# The Many-Worlds Interpretation of Quantum Mechanics

*Princeton Series in Physics*

edited by Arthur S. Wightman
and John J. Hopfield

Quantum Mechanics for Hamiltonians Defined as Quadratic Forms *by Barry Simon*

Lectures on Current Algebra and Its Applications *by Sam B. Treiman, Roman Jackiw, and David J. Gross*

Physical Cosmology *by P. J. E. Peebles*

The Many-Worlds Interpretation of Quantum Mechanics *edited by B. S. DeWitt and N. Graham*

# The Many-Worlds Interpretation of Quantum Mechanics

A Fundamental Exposition by
HUGH EVERETT, III, with Papers by
J. A. WHEELER, B. S. DeWITT,
L. N. COOPER and D. VAN VECHTEN,
and N. GRAHAM

Edited by
BRYCE S. DeWITT and NEILL GRAHAM

*Princeton Series in Physics*

Princeton University Press
Princeton, New Jersey, 1973

LC Card: 72-12116

ISBN: 0-691-08126-3 (hard cover edition)
ISBN: 0-691-88131-X (paperback edition)

Library of Congress Cataloguing in Publication data will be found on the last printed page of this book.



Printed in the United States of America
by Princeton University Press

# PREFACE

In 1957, in his Princeton doctoral dissertation, Hugh Everett, III, proposed a new interpretation of quantum mechanics that denies the existence of a separate classical realm and asserts that it makes sense to talk about a state vector for the whole universe. This state vector never collapses, and hence reality as a whole is rigorously deterministic. This reality, which is described *jointly* by the dynamical variables and the state vector, is not the reality we customarily think of, but is a reality composed of many worlds. By virtue of the temporal development of the dynamical variables the state vector decomposes naturally into orthogonal vectors, reflecting a continual splitting of the universe into a multitude of mutually unobservable but equally real worlds, in each of which every good measurement has yielded a definite result and in most of which the familiar statistical quantum laws hold.

In addition to his short thesis Everett wrote a much larger exposition of his ideas, which was never published. The present volume contains both of these works, together with a handful of papers by others on the same theme. Looked at in one way, Everett's interpretation calls for a return to naive realism and the old fashioned idea that there can be a direct correspondence between formalism and reality. Because physicists have become more sophisticated than this, and above all because the implications of his approach appear to them so bizarre, few have taken Everett seriously. Nevertheless his basic premise provides such a stimulating framework for discussions of the quantum theory of measurement that this volume should be on every quantum theoretician's shelf.

"... a picture, incomplete yet not false, of the universe as Ts'ui Pên conceived it to be. Differing from Newton and Schopenhauer,... [he] did not think of time as absolute and uniform. He believed in an infinite series of times, in a dizzily growing, ever spreading network of diverging, converging and parallel times. This web of time – the strands of which approach one another, bifurcate, intersect or ignore each other through the centuries – embraces *every* possibility. We do not exist in most of them. In some you exist and not I, while in others I do, and you do not, and in yet others both of us exist. In this one, in which chance has favored me, you have come to my gate. In another, you, crossing the garden, have found me dead. In yet another, I say these very same words, but am an error, a phantom."

Jorge Luis Borges, *The Garden of Forking Paths*

"Actualities seem to float in a wider sea of possibilities from out of which they were chosen; and *somewhere*, indeterminism says, such possibilities exist, and form part of the truth."

William James

# CONTENTS

# The Many-Worlds Interpretation of Quantum Mechanics

# THE THEORY OF THE UNIVERSAL WAVE FUNCTION

Hugh Everett, III

## I. INTRODUCTION

We begin, as a way of entering our subject, by characterizing a particular interpretation of quantum theory which, although not representative of the more careful formulations of some writers, is the most common form encountered in textbooks and university lectures on the subject.

A physical system is described completely by a state function $\psi$, which is an element of a Hilbert space, and which furthermore gives information only concerning the probabilities of the results of various observations which can be made on the system. The state function $\psi$ is thought of as objectively characterizing the physical system, i.e., at all times an isolated system is thought of as possessing a state function, independently of our state of knowledge of it. On the other hand, $\psi$ changes in a causal manner so long as the system remains isolated, obeying a differential equation. Thus there are two fundamentally different ways in which the state function can change:[1]

*Process* 1: The discontinuous change brought about by the observation of a quantity with eigenstates $\phi_1, \phi_2, \ldots,$ in which the state $\psi$ will be changed to the state $\phi_j$ with probability $|(\psi, \phi_j)|^2$.

*Process* 2: The continuous, deterministic change of state of the (isolated) system with time according to a wave equation $\frac{\partial\psi}{\partial t} = U\psi$, where $U$ is a linear operator.

---

[1] We use here the terminology of von Neumann [17].

The question of the consistency of the scheme arises if one contemplates regarding the observer and his object-system as a single (composite) physical system. Indeed, the situation becomes quite paradoxical if we allow for the existence of more than one observer. Let us consider the case of one observer A, who is performing measurements upon a system S, the totality (A + S) in turn forming the object-system for another observer, B.

If we are to deny the possibility of B's use of a quantum mechanical description (wave function obeying wave equation) for A + S, then we must be supplied with some alternative description for systems which contain observers (or measuring apparatus). Furthermore, we would have to have a criterion for telling precisely what type of systems would have the preferred positions of "measuring apparatus" or "observer" and be subject to the alternate description. Such a criterion is probably not capable of rigorous formulation.

On the other hand, if we do allow B to give a quantum description to A + S, by assigning a state function $\psi^{A+S}$, then, so long as B does not interact with A + S, its state changes causally according to Process 2, *even though* A *may be performing measurements upon* S. From B's point of view, nothing resembling Process 1 can occur (there are no discontinuities), and the question of the validity of A's use of Process 1 is raised. That is, *apparently* either A is incorrect in assuming Process 1, with its probabilistic implications, to apply to his measurements, or else B's state function, with its purely causal character, is an inadequate description of what is happening to A + S.

To better illustrate the paradoxes which can arise from strict adherence to this interpretation we consider the following amusing, but *extremely hypothetical* drama.

Isolated somewhere out in space is a room containing an observer, A, who is about to perform a measurement upon a system S. After performing his measurement he will record the result in his notebook. We assume that he knows the state function of S (perhaps as a result of previous measurement), and that it is not an eigenstate of the measurement he is about to perform. A, being an orthodox quantum theorist, then believes that the outcome of his measurement is undetermined and that the process is correctly described by Process 1.

In the meantime, however, there is another observer, B, outside the room, who is in possession of the state function of the entire room, including S, the measuring apparatus, and A, just prior to the measurement. B is only interested in what will be found in the notebook one week hence, so he computes the state function of the room for one week in the future according to Process 2. One week passes, and we find B still in possession of the state function of the room, which this equally orthodox quantum theorist believes to be a complete description of the room and its contents. If B's state function calculation tells beforehand exactly what is going to be in the notebook, then A is incorrect in his belief about the indeterminacy of the outcome of his measurement. We therefore assume that B's state function contains non-zero amplitudes over several of the notebook entries.

At this point, B opens the door to the room and looks at the notebook (performs his observation). Having observed the notebook entry, he turns to A and informs him in a patronizing manner that since his (B's) wave function just prior to his entry into the room, which he knows to have been a complete description of the room and its contents, had non-zero amplitude over other than the present result of the measurement, the result must have been decided only when B entered the room, so that A, his notebook entry, and his memory about what occurred one week ago had no independent objective existence until the intervention by B. In short, B implies that A owes his present objective existence to B's generous nature which compelled him to intervene on his behalf. However, to B's consternation, A does not react with anything like the respect and gratitude he should exhibit towards B, and at the end of a somewhat heated reply, in which A conveys in a colorful manner his opinion of B and his beliefs, he rudely punctures B's ego by observing that if B's view is correct, then he has no reason to feel complacent, since the whole present situation may have no objective existence, but may depend upon the future actions of yet another observer.

It is now clear that the interpretation of quantum mechanics with which we began is untenable if we are to consider a universe containing more than one observer. We must therefore seek a suitable modification of this scheme, or an entirely different system of interpretation. Several alternatives which avoid the paradox are:

*Alternative* 1: To postulate the existence of only one observer in the universe. This is the solipsist position, in which each of us must hold the view that he alone is the only valid observer, with the rest of the universe and its inhabitants obeying at all times Process 2 except when under his observation.

This view is quite consistent, but one must feel uneasy when, for example, writing textbooks on quantum mechanics, describing Process 1, for the consumption of other persons to whom it does not apply.

*Alternative* 2: To limit the applicability of quantum mechanics by asserting that the quantum mechanical description fails when applied to observers, or to measuring apparatus, or more generally to systems approaching macroscopic size.

If we try to limit the applicability so as to exclude measuring apparatus, or in general systems of macroscopic size, we are faced with the difficulty of sharply defining the region of validity. For what n might a group of n particles be construed as forming a measuring device so that the quantum description fails? And to draw the line at human or animal observers, i.e., to assume that all mechanical aparata obey the usual laws, but that they are somehow not valid for living observers, does violence to the so-called principle of psycho-physical parallelism,[2] and constitutes a view to be avoided, if possible. To do justice to this principle we must insist that we be able to conceive of mechanical devices (such as servomechanisms), obeying natural laws, which we would be willing to call observers.

*Alternative* 3: To admit the validity of the state function description, but to deny the possibility that B could ever be in possession of the state function of A + S. Thus one might argue that a determination of the state of A would constitute such a drastic intervention that A would cease to function as an observer.

The first objection to this view is that no matter what the state of A + S is, there is in principle a complete set of commuting operators for which it is an eigenstate, so that, at least, the determination of *these* quantities will not affect the state nor in any way disrupt the operation of A. There are no fundamental restrictions in the usual theory about the knowability of *any* state functions, and the introduction of any such restrictions to avoid the paradox must therefore require extra postulates.

The second objection is that it is not particularly relevant whether or not B actually *knows* the precise state function of A + S. If he merely *believes* that the system is described by a state function, which he does not presume to know, then the difficulty still exists. He must then believe that this state function changed deterministically, and hence that there was nothing probabilistic in A's determination.

2 In the words of von Neumann ([17], p. 418): "...it is a fundamental requirement of the scientific viewpoint – the so-called principle of the psycho-physical parallelism – that it must be possible so to describe the extra-physical process of the subjective perception as if it were in reality in the physical world – *i.e.*, to assign to its parts equivalent physical processes in the objective environment, in ordinary space."

*Alternative* 4: To abandon the position that the state function is a *complete* description of a system. The state function is to be regarded not as a description of a single system, but of an ensemble of systems, so that the probabilistic assertions arise naturally from the incompleteness of the description.

It is assumed that the correct complete description, which would presumably involve further (hidden) parameters beyond the state function alone, would lead to a deterministic theory, from which the probabilistic aspects arise as a result of our ignorance of these extra parameters in the same manner as in classical statistical mechanics.

*Alternative* 5: To assume the universal validity of the quantum description, by the complete abandonment of Process 1. The general validity of pure wave mechanics, *without any statistical assertions*, is assumed for *all* physical systems, including observers and measuring apparata. Observation processes are to be described completely by *the state function of the composite system which includes the observer and his object-system, and which at all times obeys the wave equation (Process 2).*

This brief list of alternatives is not meant to be exhaustive, but has been presented in the spirit of a preliminary orientation. We have, in fact, omitted one of the foremost interpretations of quantum theory, namely the position of Niels Bohr. The discussion will be resumed in the final chapter, when we shall be in a position to give a more adequate appraisal of the various alternate interpretations. For the present, however, we shall concern ourselves only with the development of Alternative 5.

It is evident that Alternative 5 is a theory of many advantages. It has the virtue of logical simplicity and it is complete in the sense that it is applicable to the entire universe. All processes are considered equally (there are no "measurement processes" which play any preferred role), and the principle of psycho-physical parallelism is fully maintained. Since the universal validity of the state function description is asserted, one can regard the state functions themselves as the fundamental entities, and one can even consider the state function of the whole universe. In this sense this theory can be called the theory of the "universal wave function," since all of physics is presumed to follow from this function alone. There remains, however, the question whether or not such a theory can be put into correspondence with our experience.

*The present thesis is devoted to showing that this concept of a universal wave mechanics, together with the necessary correlation machinery for its interpretation, forms a logically self consistent description of a universe in which several observers are at work.*

We shall be able to introduce into the theory systems which represent observers. Such systems can be conceived as automatically functioning machines (servomechanisms) possessing recording devices (memory) and which are capable of responding to their environment. The behavior of these observers shall always be treated within the framework of wave mechanics. Furthermore, we shall deduce the probabilistic assertions of Process 1 as *subjective* appearances to such observers, thus placing the theory in correspondence with experience. We are then led to the novel situation in which the formal theory is objectively continuous and causal, while subjectively discontinuous and probabilistic. While this point of view thus shall ultimately justify our use of the statistical assertions of the orthodox view, it enables us to do so in a logically consistent manner, allowing for the existence of other observers. At the same time it gives a deeper insight into the meaning of quantized systems, and the role played by quantum mechanical correlations.

In order to bring about this correspondence with experience for the pure wave mechanical theory, we shall exploit the correlation between subsystems of a composite system which is described by a state function. A subsystem of such a composite system does not, in general, possess an independent state function. That is, in general a composite system cannot be represented by a single pair of subsystem states, but can be represented only by a *superposition* of such pairs of subsystem states. For example, the Schrodinger wave function for a pair of particles, $\psi(x_1, x_2)$, cannot always be written in the form $\psi = \phi(x_1)\eta(x_2)$, but only in the form $\psi = \sum_{i,j} a_{ij}\phi^i(x_1)\eta^j(x_2)$. In the latter case, there is no single state for Particle 1 alone or Particle 2 alone, but only the superposition of such cases.

In fact, to any arbitrary choice of state for one subsystem there will correspond a *relative state* for the other subsystem, which will generally be dependent upon the choice of state for the first subsystem, so that the state of one subsystem is not independent, but correlated to the state of the remaining subsystem. Such correlations between systems arise from interaction of the systems, and from our point of view all measurement and observation processes are to be regarded simply as interactions between observer and object-system which produce strong correlations.

Let one regard an observer as a subsystem of the composite system: observer + object-system. It is then an inescapable consequence that after the interaction has taken place there will not, generally, exist a single observer state. There will, however, be a superposition of the composite system states, each element of which contains a definite observer state and a definite relative object-system state. Furthermore, as we shall see, *each* of these relative object-system states will be, approximately, the eigenstates of the observation corresponding to the value obtained by the observer which is described by the same element of the superposition. Thus, each element of the resulting superposition describes an observer who perceived a definite and generally different result, and to whom it appears that the object-system state has been transformed into the corresponding eigenstate. In this sense the usual assertions of Process 1 appear to hold on a subjective level to each observer described by an element of the superposition. We shall also see that correlation plays an important role in preserving consistency when several observers are present and allowed to interact with one another (to "consult" one another) as well as with other object-systems.

In order to develop a language for interpreting our pure wave mechanics for composite systems we shall find it useful to develop quantitative definitions for such notions as the "sharpness" or "definiteness" of an operator A for a state $\psi$, and the "degree of correlation" between the subsystems of a composite system or between a pair of operators in the subsystems, so that we can use these concepts in an unambiguous manner. The mathematical development of these notions will be carried out in the next chapter (II) using some concepts borrowed from Information Theory.[3] We shall develop there the general definitions of information and correlation, as well as some of their more important properties. Throughout Chapter II we shall use the language of probability theory to facilitate the exposition, and because it enables us to introduce in a unified manner a number of concepts that will be of later use. We shall nevertheless subsequently apply the mathematical definitions directly to state functions, by replacing probabilities by square amplitudes, *without*, however, *making any reference to probability models.*

Having set the stage, so to speak, with Chapter II, we turn to quantum mechanics in Chapter III. There we first investigate the quantum formalism of composite systems, particularly the concept of relative state functions, and the meaning of the representation of subsystems by non-interfering mixtures of states characterized by density matrices. The notions of information and correlation are then applied to quantum mechanics. The final section of this chapter discusses the measurement process, which is regarded simply as a correlation-inducing interaction between subsystems of a single isolated system. A simple example of such a measurement is given and discussed, and some general consequences of the superposition principle are considered.

---

3 The theory originated by Claude E. Shannon [19].

This will be followed by an abstract treatment of the problem of Observation (Chapter IV). In this chapter we make use only of the superposition principle, and general rules by which composite system states are formed of subsystem states, in order that our results shall have the greatest generality and be applicable to any form of quantum theory for which these principles hold. (Elsewhere, when giving examples, we restrict ourselves to the non-relativistic Schrödinger Theory for simplicity.) The validity of Process 1 as a subjective phenomenon is deduced, as well as the consistency of allowing several observers to interact with one another.

Chapter V supplements the abstract treatment of Chapter IV by discussing a number of diverse topics from the point of view of the theory of pure wave mechanics, including the existence and meaning of macroscopic objects in the light of their atomic constitution, amplification processes in measurement, questions of reversibility and irreversibility, and approximate measurement.

The final chapter summarizes the situation, and continues the discussion of alternate interpretations of quantum mechanics.

## II. PROBABILITY, INFORMATION, AND CORRELATION

The present chapter is devoted to the mathematical development of the concepts of information and correlation. As mentioned in the introduction we shall use the language of probability theory throughout this chapter to facilitate the exposition, although we shall apply the mathematical definitions and formulas in later chapters without reference to probability models. We shall develop our definitions and theorems in full generality, for probability distributions over arbitrary sets, rather than merely for distributions over real numbers, with which we are mainly interested at present. We take this course because it is as easy as the restricted development, and because it gives a better insight into the subject.

The first three sections develop definitions and properties of information and correlation for probability distributions over *finite* sets only. In section four the definition of correlation is extended to distributions over arbitrary sets, and the general invariance of the correlation is proved. Section five then generalizes the definition of information to distributions over arbitrary sets. Finally, as illustrative examples, sections seven and eight give brief applications to stochastic processes and classical mechanics, respectively.

### §1. *Finite joint distributions*

We assume that we have a collection of finite sets, $\mathfrak{X},\mathfrak{Y},\ldots,\mathfrak{Z}$, whose elements are denoted by $x_i \in \mathfrak{X}$, $y_j \in \mathfrak{Y},\ldots,$ $z_k \in \mathfrak{Z}$, etc., and that we have a *joint probability distribution*, $P = P(x_i,y_j,\ldots,z_k)$, defined on the cartesian product of the sets, which represents the probability of the combined event $x_i,y_j,\ldots,$ and $z_k$. We then denote by $X,Y,\ldots,Z$ the random variables whose values are the elements of the sets $\mathfrak{X},\mathfrak{Y},\ldots,\mathfrak{Z}$, with probabilities given by $P$.

For any subset $Y,\ldots,Z$, of a set of random variables $W,\ldots,X, Y,\ldots,Z$, with joint probability distribution $P(w_i,\ldots,x_j,y_k,\ldots,z_\ell)$, the *marginal distribution*, $P(y_k,\ldots,z_\ell)$, is defined to be:

$$P(y_k,\ldots,z_\ell) = \sum_{i,\ldots,j} P(w_i,\ldots,x_j,y_k,\ldots,z_\ell) \ , \tag{1.1}$$

which represents the probability of the joint occurrence of $y_k,\ldots,z_\ell$, with no restrictions upon the remaining variables.

For any subset $Y,\ldots,Z$ of a set of random variables the *conditional distribution*, conditioned upon the values $W = w_i,\ldots,X = x_j$ for any remaining subset $W,\ldots,X$, and denoted by $P^{w_i,\ldots,x_j}(y_k,\ldots,z_\ell)$, is defined to be:[1]

$$P^{w_i,\ldots,x_j}(y_k,\ldots,z_\ell) = \frac{P(w_i,\ldots,x_j,y_k,\ldots,z_\ell)}{P(w_i,\ldots,x_j)} \ , \tag{1.2}$$

which represents the probability of the joint event $Y = y_k,\ldots,Z = z_\ell$, conditioned by the fact that $W,\ldots,X$ are known to have taken the values $w_i,\ldots,x_j$, respectively.

For any numerical valued function $F(y_k,\ldots,z_\ell)$, defined on the elements of the cartesian product of $\mathcal{Y},\ldots,\mathcal{Z}$, the *expectation*, denoted by $\text{Exp}\,[F]$, is defined to be:

$$\text{Exp}\,[F] = \sum_{k,\ldots,\ell} P(y_k,\ldots,z_\ell)\, F(y_k,\ldots,z_\ell) \ . \tag{1.3}$$

We note that if $P(y_k,\ldots,z_\ell)$ is a marginal distribution of some larger distribution $P(w_i,\ldots,x_j,y_k,\ldots,z_\ell)$ then

$$\begin{aligned} \text{Exp}\,[F] &= \sum_{k,\ldots,\ell} \left( \sum_{i,\ldots,j} P(w_i,\ldots,x_j,y_k,\ldots,z_\ell) \right) F(y_k,\ldots,z_\ell) \\ &= \sum_{i,\ldots,j,k,\ldots,\ell} P(w_i,\ldots,x_j,y_k,\ldots,z_\ell)\, F(y_k,\ldots,z_\ell) \ , \end{aligned} \tag{1.4}$$

---

1 We regard it as undefined if $P(w_i,\ldots,x_j) = 0$. In this case $P(w_i,\ldots,x_j, y_k,\ldots,z_\ell)$ is necessarily zero also.

so that if we wish to compute Exp [F] with respect to some joint distribution it suffices to use *any* marginal distribution of the original distribution which contains at least those variables which occur in F.

We shall also occasionally be interested in *conditional expectations*, which we define as:

$$\text{Exp}^{w_i,\ldots,x_j}[F] = \sum_{k,\ldots,\ell} P^{w_i,\ldots,x_j}(y_k,\ldots,z_\ell)F(y_k,\ldots,z_\ell) \ , \tag{1.5}$$

and we note the following easily verified rules for expectations:

$$\text{Exp}\,[\text{Exp}\,[F]] = \text{Exp}\,[F] \ , \tag{1.6}$$

$$\text{Exp}^{u_i,\ldots,v_j}[\text{Exp}^{u_i,\ldots,v_j,\,w_k,\ldots,x_\ell}[F]] = \text{Exp}^{u_i,\ldots,v_j}[F] \ , \tag{1.7}$$

$$\text{Exp}\,[F+G] = \text{Exp}\,[F] + \text{Exp}\,[G] \ . \tag{1.8}$$

We should like finally to comment upon the notion of *independence*. Two random variables X and Y with joint distribution $P(x_i, y_j)$ will be said to be independent if and only if $P(x_i, y_j)$ is equal to $P(x_i)P(y_j)$ for all i, j. Similarly, the groups of random variables (U...V), (W...X),..., (Y...Z) will be called *mutually independent groups* if and only if $P(u_i,\ldots,v_j, w_k,\ldots,x_\ell,\ldots,y_m,\ldots,z_n)$ is always equal to $P(u_i,\ldots,v_j)$ $P(w_k,\ldots,x_\ell)\ldots P(y_m,\ldots,z_n)$.

Independence means that the random variables take on values which are not influenced by the values of other variables with respect to which they are independent. That is, the conditional distribution of one of two independent variables, Y, conditioned upon the value $x_i$ for the other, is independent of $x_i$, so that knowledge about one variable tells nothing of the other.

## §2. *Information for finite distributions*

Suppose that we have a single random variable X, with distribution $P(x_i)$. We then define[2] a number, $I_X$, called the *information* of X, to be:

[2] This definition corresponds to the negative of the *entropy* of a probability distribution as defined by Shannon [19].

(2.1) $$I_X = \sum_i P(x_i) \ln P(x_i) = \text{Exp}\,[\ln P(x_i)]\ ,$$

which is a function of the probabilities alone and not of any possible numerical values of the $x_i$'s themselves.[3]

The information is essentially a measure of the sharpness of a probability distribution, that is, an inverse measure of its "spread." In this respect information plays a role similar to that of variance. However, it has a number of properties which make it a superior measure of the "sharpness" than the variance, not the least of which is the fact that it can be defined for distributions over arbitrary sets, while variance is defined only for distributions over real numbers.

Any change in the distribution $P(x_i)$ which "levels out" the probabilities decreases the information. It has the value zero for "perfectly sharp" distributions, in which the probability is one for one of the $x_i$ and zero for all others, and ranges downward to $-\ln n$ for distributions over $n$ elements which are equal over all of the $x_i$. The fact that the information is nonpositive is no liability, since we are seldom interested in the absolute information of a distribution, but only in differences.

We can generalize (2.1) to obtain the formula for the information of a group of random variables $X, Y, \ldots, Z$, with joint distribution $P(x_i, y_j, \ldots, z_k)$, which we denote by $I_{XY\ldots Z}$:

(2.2) $$I_{XY\ldots Z} = \sum_{i,j,\ldots,k} P(x_i, y_j, \ldots, z_k) \ln P(x_i, y_j, \ldots, z_k)$$
$$= \text{Exp}\,[\ln P(x_i, y_j, \ldots, z_k)]\ ,$$

3 A good discussion of information is to be found in Shannon [19], or Woodward [21]. Note, however, that in the theory of communication one defines the information of a *state* $x_i$, which has *a priori* probability $P_i$, to be $-\ln P_i$. We prefer, however, to regard information as a property of the distribution itself.

which follows immediately from our previous definition, since the group of random variables $X, Y, \ldots, Z$ may be regarded as a single random variable $W$ which takes its values in the cartesian product $\mathcal{X} \times \mathcal{Y} \times \cdots \times \mathcal{Z}$.

Finally, we define a *conditional information*, $I_{XY\ldots Z}^{v_m,\ldots,w_n}$, to be:

$$(2.3)\quad I_{XY\ldots Z}^{v_m,\ldots,w_n} = \sum_{i,j,\ldots,k} P^{v_m,\ldots,w_n}(x_i, y_j,\ldots,z_k) \ln P^{v_m,\ldots,w_n}(x_i, y_j,\ldots,z_k)$$
$$= \mathrm{Exp}^{v_m,\ldots,w_n}\left[\ln P^{v_m,\ldots,w_n}(x_i, y_j,\ldots,z_k)\right] ,$$

a quantity which measures our information about $X, Y, \ldots, Z$ given that we know that $V\ldots W$ have taken the particular values $v_m, \ldots, w_n$.

For independent random variables $X, Y, \ldots, Z$, the following relationship is easily proved:

$$(2.4)\qquad I_{XY\ldots Z} = I_X + I_Y + \ldots + I_Z \quad (X, Y, \ldots, Z \text{ independent}) ,$$

so that the information of $XY\ldots Z$ is the sum of the individual quantities of information, which is in accord with our intuitive feeling that if we are given information about unrelated events, our total knowledge is the sum of the separate amounts of information. We shall generalize this definition later, in §5.

## §3. *Correlation for finite distributions*

Suppose that we have a pair of random variables, $X$ and $Y$, with joint distribution $P(x_i, y_j)$. If we say that $X$ and $Y$ are *correlated*, what we intuitively mean is that *one learns something about one variable when he is told the value of the other*. Let us focus our attention upon the variable $X$. If we are not informed of the value of $Y$, then our information concerning $X$, $I_X$, is calculated from the marginal distribution $P(x_i)$. However, if we are now told that $Y$ has the value $y_j$, then our information about $X$ changes to the information of the conditional distribution $P^{y_j}(x_i)$, $I_X^{y_j}$. According to what we have said, we wish the degree correlation to measure how much we learn about $X$ by being informed of

Y's value. However, since the change of information, $I_X^{y_j} - I_X$, may depend upon the particular value, $y_j$, of Y which we are told, the natural thing to do to arrive at a single number to measure the strength of correlation is to consider the *expected* change in information about X, given that we are to be told the value of Y. This quantity we call the *correlation information*, or for brevity, the *correlation*, of X and Y, and denote it by $\{X, Y\}$. Thus:

$$\{X, Y\} = \mathrm{Exp}\left[I_X^{y_j} - I_X\right] = \mathrm{Exp}\left[I_X^{y_j}\right] - I_X \ . \tag{3.1}$$

Expanding the quantity $\mathrm{Exp}\left[I_X^{y_j}\right]$ using (2.3) and the rules for expectations (1.6)–(1.8) we find:

$$\mathrm{Exp}\left[I_X^{y_j}\right] = \mathrm{Exp}\left[\mathrm{Exp}^{y_j}\left[\ln P^{y_j}(x_i)\right]\right]$$

$$= \mathrm{Exp}\left[\ln \frac{P(x_i, y_j)}{P(y_j)}\right] = \mathrm{Exp}\left[\ln P(x_i, y_j)\right] - \mathrm{Exp}\left[\ln P(y_j)\right] \tag{3.2}$$

$$= I_{XY} - I_Y \ ,$$

and combining with (3.1) we have:

$$\{X, Y\} = I_{XY} - I_X - I_Y \ . \tag{3.3}$$

Thus the correlation is symmetric between X and Y, and hence also equal to the expected change of information about Y given that we will be told the value of X. Furthermore, according to (3.3) the correlation corresponds precisely to the amount of "missing information" if we possess only the marginal distributions, i.e., the loss of information if we choose to regard the variables as independent.

THEOREM 1. $\{X, Y\} = 0$ *if and only if* X *and* Y *are independent, and is otherwise strictly positive.* (Proof in Appendix I.)

In this respect the correlation so defined is superior to the usual correlation coefficients of statistics, such as covariance, etc., which can be zero even when the variables are not independent, and which can assume both positive and negative values. An inverse correlation is, after all, quite as useful as a direct correlation. Furthermore, it has the great advantage of depending upon the probabilities alone, and not upon any numerical values of $x_i$ and $y_j$, so that it is defined for distributions over sets whose elements are of an arbitrary nature, and not only for distributions over numerical properties. For example, we might have a joint probability distribution for the political party and religious affiliation of individuals. Correlation and information are defined for such distributions, although they possess nothing like covariance or variance.

We can generalize (3.3) to define a *group correlation* for the groups of random variables $(U\ldots V), (W\ldots X), \ldots, (Y\ldots Z)$, denoted by $\{U\ldots V, W\ldots X, \ldots, Y\ldots Z\}$ (where the groups are separated by commas), to be:

$$\{U\ldots V, W\ldots X, \ldots, Y\ldots Z\} = I_{U\ldots V W\ldots X\ldots Y\ldots Z} - I_{U\ldots V} - I_{W\ldots X} - \cdots - I_{Y\ldots Z}, \tag{3.4}$$

again measuring the information deficiency for the group marginals. Theorem 1 is also satisfied by the group correlation, so that it is zero if and only if the groups are mutually independent. We can, of course, also define conditional correlations in the obvious manner, denoting these quantities by appending the conditional values as superscripts, as before.

We conclude this section by listing some useful formulas and inequalities which are easily proved:

$$\{U, V, \ldots, W\} = \mathrm{Exp}\left[\ln \frac{P(u_i, v_j, \ldots, w_k)}{P(u_i)P(v_j)\ldots P(w_k)}\right], \tag{3.5}$$

$$\{U,V,\ldots,W\}^{x_i\ldots y_j} = \mathrm{Exp}^{x_i\ldots y_j}\left[\ln \frac{P^{x_i\ldots y_j}(u_k, v_1, \ldots, w_m)}{P^{x_i\ldots y_j}(u_k)P^{x_i\ldots y_j}(v_1)\ldots P^{x_i\ldots y_j}(w_m)}\right] \tag{3.6}$$

(conditional correlation) ,

$$(3.7)\quad \begin{aligned} &\{\ldots,U,V,\ldots\} = \{\ldots,UV,\ldots\} + \{U,V\}\ , \\ &\{\ldots,U,V,\ldots,W,\ldots\} = \{\ldots,UV\ldots W,\ldots\} + \{U,V,\ldots,W\} \text{ (comma removal)} \end{aligned}$$

$$(3.8)\quad \{\ldots,U,VW,\ldots\} - \{\ldots,UV,W,\ldots\} = \{U,V\} - \{V,W\} \text{ (commutator)}\ ,$$

$$(3.9)\quad \{X\} = 0 \quad \text{(definition of bracket with no commas)}\ ,$$

$$(3.10)\quad \{\ldots,XXV,\ldots\} = \{\ldots,XV,\ldots\}$$

(removal of repeated variable within a group) ,

$$(3.11)\quad \{\ldots,UV,VW,\ldots\} = \{\ldots,UV,W,\ldots\} - \{V,W\} - I_V$$

(removal of repeated variable in separate groups) ,

$$(3.12)\quad \{X,X\} = -I_X \quad \text{(self correlation)}\ ,$$

$$(3.13)\quad \begin{aligned} &\{U,VW,X\}^{\ldots w_j \ldots} = \{U,V,X\}^{\ldots w_j \ldots}\ , \\ &\{U,W,X\}^{\ldots w_j \ldots} = \{U,X\}^{\ldots w_j \ldots} \end{aligned}$$

(removal of conditioned variables) ,

$$(3.14)\quad \{XY,Z\} \geqq \{X,Z\}\ ,$$

$$(3.15)\quad \{XY,Z\} \geqq \{X,Z\} + \{Y,Z\} - \{X,Y\}\ ,$$

$$(3.16)\quad \{X,Y,Z\} \geqq \{X,Y\} + \{X,Z\}\ .$$

Note that in the above formulas any random variable W may be replaced by any group XY...Z and the relation holds true, since the set XY...Z may be regarded as the single random variable W, which takes its values in the cartesian product $\mathfrak{X} \times \mathfrak{Y} \times \ldots \times \mathfrak{Z}$.

## §4. *Generalization and further properties of correlation*

Until now we have been concerned only with finite probability distributions, for which we have defined information and correlation. We shall now generalize the definition of correlation so as to be applicable to joint probability distributions over arbitrary sets of unrestricted cardinality.

We first consider the effects of refinement of a finite distribution. For example, we may discover that the event $x_i$ is actually the disjunction of several exclusive events $\tilde{x}_i^1,\ldots,\tilde{x}_i^n$, so that $x_i$ occurs if any one of the $\tilde{x}_i^\mu$ occurs, i.e., the single event $x_i$ results from failing to distinguish between the $\tilde{x}_i^\mu$. The probability distribution which distinguishes between the $\tilde{x}_i^\mu$ will be called a *refinement* of the distribution which does not. In general, we shall say that a distribution $P' = P'(\tilde{x}_i^\mu,\ldots,\tilde{y}_j^\nu)$ is a refinement of $P = P(x_i,\ldots,y_j)$ if

$$P(x_i,\ldots,y_j) = \sum_{\mu\ldots\nu} P'(\tilde{x}_i^\mu,\ldots,\tilde{y}_j^\nu) \quad (\text{all } i,\ldots,j)\ . \tag{4.1}$$

We now state an important theorem concerning the behavior of correlation under a refinement of a joint probability distributions:

THEOREM 2. *$P'$ is a refinement of* $P \Rightarrow \{X,\ldots,Y\}' \geq \{X,\ldots,Y\}$ *so that correlations never decrease upon refinement of a distribution.* (Proof in Appendix I, §3.)

As an example, suppose that we have a continuous probability density $P(x, y)$. By division of the axes into a finite number of intervals, $\bar{x}_i$, $\bar{y}_j$, we arrive at a finite joint distribution $P_{ij}$, by integration of $P(x, y)$ over the rectangle whose sides are the intervals $\bar{x}_i$ and $\bar{y}_j$, and which represents the probability that $X \epsilon \bar{x}_i$ and $Y \epsilon \bar{y}_j$. If we now subdivide the intervals, the new distribution $P'$ will be a refinement of $P$, and by Theorem 2 the correlation $\{X,Y\}$ computed from $P'$ will never be less than that computed from $P$. Theorem 2 is seen to be simply the mathematical verification of the intuitive notion that closer analysis of a situation in which quantities $X$ and $Y$ are dependent can never lessen the knowledge about $Y$ which can be obtained from $X$.

This theorem allows us to give a general definition of correlation which will apply to joint distributions over completely arbitrary sets, i.e., for any probability measure[4] on an arbitrary product space, in the following manner:

Assume that we have a collection of arbitrary sets $\mathcal{X}, \mathcal{Y}, \ldots, \mathcal{Z}$, and a probability measure, $M_P(\mathcal{X} \times \mathcal{Y} \times \cdots \times \mathcal{Z})$, on their cartesian product. Let $\mathcal{P}^\mu$ be any finite partition of $\mathcal{X}$ into subsets $\mathcal{X}_i^\mu$, $\mathcal{Y}$ into subsets $\mathcal{Y}_j^\mu, \ldots$, and $\mathcal{Z}$ into subsets $\mathcal{Z}_k^\mu$, such that the sets $\mathcal{X}_i^\mu \times \mathcal{Y}_j^\mu \times \ldots \times \mathcal{Z}_k^\mu$ of the cartesian product are measurable in the probability measure $M_P$. Another partition $\mathcal{P}^\nu$ is a *refinement* of $\mathcal{P}^\mu$, $\mathcal{P}^\nu \subseteq \mathcal{P}^\mu$, if $\mathcal{P}^\nu$ results from $\mathcal{P}^\mu$ by further subdivision of the subsets $\mathcal{X}_i^\mu, \mathcal{Y}_j^\mu, \ldots, \mathcal{Z}_k^\mu$. Each partition $\mathcal{P}^\mu$ results in a finite probability distribution, for which the correlation, $\{X, Y, \ldots, Z\}^{\mathcal{P}^\mu}$, is always defined through (3.3). Furthermore a refinement of a partition leads to a refinement of the probability distribution, so that by Theorem 2:

$$\mathcal{P}^\nu \subseteq \mathcal{P}^\mu \Rightarrow \{X, Y, \ldots, Z\}^{\mathcal{P}^\nu} \geq \{X, Y, \ldots, Z\}^{\mathcal{P}^\mu} \ . \tag{4.8}$$

Now the set of all partitions is partially ordered under the refinement relation. Moreover, because for any pair of partitions $\mathcal{P}, \mathcal{P}'$ there is always a third partition $\mathcal{P}''$ which is a refinement of both (common lower bound), the set of all partitions forms a *directed set*.[5] For a function, f, on a directed set, $\mathcal{S}$, one defines a directed set limit, lim f,:

**DEFINITION.** lim f exists and is equal to $a \Leftrightarrow$ for every $\varepsilon > 0$ there exists an $\alpha \in \mathcal{S}$ such that $|f(\beta) - a| < \varepsilon$ for every $\beta \in \mathcal{S}$ for which $\beta \leq \alpha$.

It is easily seen from the directed set property of common lower bounds that if this limit exists it is necessarily unique.

4 A measure is a non-negative, countably additive set function, defined on some subsets of a given set. It is a probability measure if the measure of the entire set is unity. See Halmos [12].

5 See Kelley [15], p. 65.

By (4.8) the correlation $\{X,Y,\ldots,Z\}^{\mathcal{P}}$ is a *monotone* function on the directed set of all partitions. Consequently the directed set limit, which we shall take as the basic definition of the correlation $\{X,Y,\ldots,Z\}$, *always exists.* (It may be infinite, but it is in every case well defined.) Thus:

DEFINITION. $\{X,Y,\ldots,Z\} = \lim \{X,Y,\ldots,Z\}^{\mathcal{P}}$ ,

and we have succeeded in our endeavor to give a completely general definition of correlation, applicable to all types of distributions.

It is an immediate consequence of (4.8) that this directed set limit is the supremum of $\{X,Y,\ldots,Z\}^{\mathcal{P}}$, so that:

$$\{X,Y,\ldots,Z\} = \sup_{\mathcal{P}} \{X,Y,\ldots,Z\}^{\mathcal{P}} , \tag{4.9}$$

which we could equally well have taken as the definition.

Due to the fact that the correlation is defined as a limit for discrete distributions, Theorem 1 and all of the relations (3.7) to (3.15), which contain only correlation brackets, remain true for arbitrary distributions. Only (3.11) and (3.12), which contain information terms, cannot be extended.

We can now prove an important theorem about correlation which concerns its invariant nature. Let $\mathfrak{X}, \mathfrak{Y},\ldots, \mathfrak{Z}$ be arbitrary sets with probability measure $M_P$ on their cartesian product. Let f be any one-one mapping of $\mathfrak{X}$ onto a set $\mathfrak{U}$, g a one-one map of $\mathfrak{Y}$ onto $\mathfrak{V},\ldots,$ and h a map of $\mathfrak{Z}$ onto $\mathfrak{W}$. Then a joint probability distribution over $\mathfrak{X}\times\mathfrak{Y}\times\cdots\times\mathfrak{Z}$ leads also to one over $\mathfrak{U}\times\mathfrak{V}\times\cdots\times\mathfrak{W}$ where the probability $M'_P$ induced on the product $\mathfrak{U}\times\mathfrak{V}\times\cdots\times\mathfrak{W}$ is simply the measure which assigns to each subset of $\mathfrak{U}\times\mathfrak{V}\times\cdots\times\mathfrak{W}$ the measure which is the measure of its image set in $\mathfrak{X}\times\mathfrak{Y}\times\cdots\times\mathfrak{Z}$ for the original measure $M_P$. (We have simply transformed to a new set of random variables: $U = f(X)$, $V = g(Y)$, $\ldots$, $W = h(Z)$.) Consider any partition $\mathcal{P}$ of $\mathfrak{X}, \mathfrak{Y},\ldots, \mathfrak{Z}$ into the subsets $\{\mathfrak{X}_i\}, \{\mathfrak{Y}_j\},\ldots, \{\mathfrak{Z}_k\}$ with probability distribution $P_{ij\ldots k} = M_P(\mathfrak{X}_i\times\mathfrak{Y}_j\times,\ldots,\times\mathfrak{Z}_k)$. Then there is a corresponding partition $\mathcal{P}'$ of $\mathfrak{U}, \mathfrak{V},\ldots, \mathfrak{W}$ into the image sets of the sets of $\mathcal{P}, \{\mathcal{U}_i\}, \{\mathcal{V}_j\}, \ldots, \{\mathcal{W}_k\}$, where $\mathcal{U}_i = f(\mathcal{X}_i)$, $\mathcal{V}_j = g(\mathcal{Y}_j), \ldots,$ $\mathcal{W}_k = h(\mathcal{Z}_k)$. But the probability distribution for $\mathcal{P}'$ is the same as that for $\mathcal{P}$, since $P'_{ij\ldots k} = M'_P(\mathcal{U}_i \times \mathcal{V}_j \times \cdots \times \mathcal{W}_k) = M_P(\mathcal{X}_i \times \mathcal{Y}_j \times \cdots \times \mathcal{Z}_k) = P_{ij\ldots k}$, so that:

$$\{X,Y,\ldots,Z\}^{\mathcal{P}} = \{U,V,\ldots,W\}^{\mathcal{P}'} . \tag{4.10}$$

Due to the correspondence between the $\mathcal{P}$'s and $\mathcal{P}'$'s we have that:

$$\sup_{\mathcal{P}} \{X,Y,\ldots,Z\}^{\mathcal{P}} = \sup_{\mathcal{P}'} \{U,V,\ldots,W\}^{\mathcal{P}'} , \tag{4.11}$$

and by virtue of (4.9) we have proved the following theorem:

THEOREM 3. $\{X,Y,\ldots,Z\} = \{U,V,\ldots,W\}$, *where* $\mathcal{U}, \mathcal{V}, \ldots, \mathcal{W}$ *are any one-one images of* $\mathcal{X}, \mathcal{Y}, \ldots, \mathcal{Z}$, *respectively. In other notation:* $\{X,Y,\ldots,Z\} = \{f(X), g(Y), \ldots, h(Z)\}$ *for all one-one functions* f, g,..., h.

This means that changing variables to functionally related variables preserves the correlation. Again this is plausible on intuitive grounds, since a knowledge of $f(x)$ is just as good as knowledge of $x$, provided that $f$ is one-one.

A special consequence of Theorem 3 is that for any continuous probability density $P(x,y)$ over real numbers the correlation between $f(x)$ and $g(y)$ is the same as between $x$ and $y$, where $f$ and $g$ are any real valued one-one functions. As an example consider a probability distribution for the position of two particles, so that the random variables are the position coordinates. Theorem 3 then assures us that the position correlation is *independent of the coordinate system*, even if different coordinate systems are used for each particle! Also for a joint distribution for a pair of events in space-time the correlation is invariant to arbitrary space-time coordinate transformations, again even allowing different transformations for the coordinates of each event.

These examples illustrate clearly the *intrinsic nature* of the correlation of various groups for joint probability distributions, which is implied by its invariance against arbitrary (one-one) transformations of the random variables. These correlation quantities are thus fundamental properties of probability distributions. A correlation is an *absolute* rather than *relative* quantity, in the sense that the correlation between (numerical valued) random variables is completely independent of the scale of measurement chosen for the variables.

## §5. *Information for general distributions*

Although we now have a definition of correlation applicable to all probability distributions, we have not yet extended the definition of information past finite distributions. In order to make this extension we first generalize the definition that we gave for discrete distributions to a definition of *relative* information for a random variable, relative to a given underlying measure, called the *information measure*, on the values of the random variable.

If we assign a measure to the set of values of a random variable, X, which is simply the assignment of a positive number $a_i$ to each value $x_i$ in the finite case, we define the information of a probability distribution $P(x_i)$ *relative* to this *information measure* to be:

$$(5.1) \qquad I_X = \sum_i P(x_i) \ln \frac{P(x_i)}{a_i} = \mathrm{Exp}\left[\ln \frac{P(x_i)}{a_i}\right].$$

If we have a joint distribution of random variables $X,Y,\ldots,Z$, with information measures $\{a_i\},\{b_j\},\ldots,\{c_k\}$ on their values, then we define the total information relative to these measures to be:

$$(5.2) \qquad I_{XY\ldots Z} = \sum_{ij\ldots k} P(x_i,y_j,\ldots,z_k) \ln \frac{P(x_i,y_j,\ldots,z_k)}{a_i b_j \ldots c_k}$$

$$= \mathrm{Exp}\left[\ln \frac{P(x_i,y_j,\ldots,z_k)}{a_i b_j \ldots c_k}\right],$$

so that the information measure on the cartesian product set is *always* taken to be the product measure of the individual information measures.

We shall now alter our previous position slightly and consider information as always being defined relative to some information measure, so that our previous definition of information is to be regarded as the information relative to the measure for which all the $a_i$'s, $b_j$'s,... and $c_k$'s are taken to be unity, which we shall henceforth call the *uniform measure.*

Let us now compute the correlation $\{X,Y,...,Z\}'$ by (3.4) using the relative information:

$$(5.3)\quad \{X,Y,...,Z\}' = I'_{XY...Z} - I'_X - I'_Y - \dots - I'_Z$$

$$= \text{Exp}\left[\ln \frac{P(x_i,y_j,...,z_k)}{a_i b_j \dots c_k}\right] - \text{Exp}\left[\ln \frac{P(x_i)}{a_i}\right] - \dots - $$

$$\text{Exp}\left[\ln \frac{P(s_k)}{c_k}\right]$$

$$= \text{Exp}\left[\ln \frac{P(x_i,y_j,...,z_k)}{P(x_i)P(y_j)\dots P(z_k)}\right] = \{X,Y,...,Z\}\ ,$$

so that the correlation for discrete distributions, as defined by (3.4), is independent of the choice of information measure, and the correlation remains an absolute, not relative quantity. It can, however, be computed from the information relative to any information measure through (3.4).

If we consider refinements, of our distributions, as before, and realize that such a refinement is also a refinement of the information measure, then we can prove a relation analogous to Theorem 2:

THEOREM 4. *The information of a distribution relative to a given information measure never decreases under refinement.* (Proof in Appendix I.)

Therefore, just as for correlation, we can define the information of a probability measure $M_P$ on the cartesian product of arbitrary sets

$\mathcal{X}, \mathcal{Y}, \ldots, \mathcal{Z}$, relative to the information measures $\mu_X, \mu_Y, \ldots, \mu_Z$, on the individual sets, by considering finite partitions $\mathcal{P}$ into subsets $\{\mathcal{X}_i\}$, $\{\mathcal{Y}_j\}, \ldots, \{\mathcal{Z}_k\}$, for which we take as the definition of the information:

$$(5.4) \qquad I^{\mathcal{P}}_{XY\ldots Z} = \sum_{ij\ldots k} M_P(\mathcal{X}_i, \mathcal{Y}_j, \ldots, \mathcal{Z}_k) \ln \frac{M_P(\mathcal{X}_i, \mathcal{Y}_j, \ldots, \mathcal{Z}_k)}{\mu_X(\mathcal{X}_i)\mu_Y(\mathcal{Y}_j)\ldots\mu_Z(\mathcal{Z}_k)} .$$

Then $I^{\mathcal{P}}_{XY\ldots Z}$ is, as was $\{X, Y, \ldots, Z\}^{\mathcal{P}}$, a monotone function upon the directed set of partitions (by Theorem 4), and as before we take the directed set limit for our definition:

$$(5.5) \qquad I_{XY\ldots Z} = \lim I^{\mathcal{P}}_{XY\ldots Z} = \sup_{\mathcal{P}} I^{\mathcal{P}}_{XY\ldots Z}$$

which is then the information relative to the information measures $\mu_X, \mu_Y, \ldots, \mu_Z$.

Now, for functions f, g on a directed set the existence of lim f and lim g is a sufficient condition for the existence of lim(f+g), which is then lim f + lim g, provided that this is not indeterminate. Therefore:

THEOREM 5. $\{X, \ldots, Y\} = \lim \{X, \ldots, Y\}^{\mathcal{P}} = \lim \left[I^{\mathcal{P}}_{X\ldots Y} - I^{\mathcal{P}}_X - \ldots - I^{\mathcal{P}}_Y\right] = I_{X\ldots Y} - I_X - \ldots - I_Y$, *where the information is taken relative to any information measure for which the expression is not indeterminate. It is sufficient for the validity of the above expression that the basic measures* $\mu_X, \ldots, \mu_Y$ *be such that none of the marginal informations* $I_X \ldots I_Y$ *shall be positively infinite.*

The latter statement holds since, because of the general relation $I_{X\ldots Y} \geq I_X + \ldots + I_Y$, the determinateness of the expression is guaranteed so long as all of the $I_X, \ldots, I_Y$ are $< +\infty$.

Henceforth, unless otherwise noted, we shall understand that information is to be computed with respect to the uniform measure for discrete distributions, and Lebesgue measure for continuous distributions over real numbers. In case of a mixed distribution, with a continuous density $P(x,y,\ldots,z)$ plus discrete "lumps" $P'(x_i,y_j,\ldots,z_k)$, we shall understand the information measure to be the uniform measure over the discrete range, and Lebesgue measure over the continuous range. These conventions then lead us to the expressions:

$$(5.6)\qquad I_{XY\ldots Z} = \begin{cases} \sum\limits_{ij\ldots k} P(x_i,y_j,\ldots,z_k) \ln P(x_i,y_j,\ldots,z_k) & \text{(discrete)} \\ \int P(x,y,\ldots,z) \ln P(x,y,\ldots,z)\, dx dy \ldots dz & \text{(cont.)} \\ \left.\begin{array}{l} \sum\limits_{i\ldots k} P'(x_i,\ldots,z_k) \ln P(x_i,\ldots,z_k) \\ + \int P(x,\ldots,z) \ln P(x,\ldots,z)\, dx\ldots dz \end{array}\right\} & \text{(mixed)} \end{cases}$$

(unless otherwise noted) .

The mixed case occurs often in quantum mechanics, for quantities which have both a discrete and continuous spectrum.

§6. *Example: Information decay in stochastic processes*

As an example illustrating the usefulness of the concept of relative information we shall consider briefly stochastic processes.[6] Suppose that we have a stationary Markov[7] process with a finite number of states $S_i$, and that the process occurs at discrete (integral) times $1,2,\ldots,n,\ldots$, at which times the transition probability from the state $S_i$ to the state $S_j$ is $T_{ij}$. The probabilities $T_{ij}$ then form what is called a *stochastic*

---

6 See Feller [10], or Doob [6].

7 A Markov process is a stochastic process whose future development depends only upon its present state, and not on its past history.

*matrix*, i.e., the elements are between 0 and 1, and $\sum_j T_{ij} = 1$ for all i. If at any time k the probability distribution over the states is $\{P_i^k\}$ then at the next time the probabilities will be $P_j^{k+1} = \sum_i P_i^k T_{ij}$.

In the special case where the matrix is *doubly-stochastic*, which means that $\sum_i T_{ij}$, as well as $\sum_j T_{ij}$, equals unity, and which amounts to a principle of detailed balancing holding, it is known that the entropy of a probability distribution over the states, defined as $H = -\sum_i P_i \ln P_i$, is a monotone increasing function of the time. This entropy is, however, simply the negative of the information relative to the uniform measure.

One can extend this result to more general stochastic processes only if one uses the more general definition of relative infomation. For an arbitrary stationary process the choice of an information measure which is stationary, i.e., for which

$$a_j = \sum_i a_i T_{ij} \quad \text{(all } j) \tag{6.1}$$

leads to the desired result. In this case the *relative* information,

$$I = \sum_i P_i \ln \frac{P_i}{a_i} , \tag{6.2}$$

is a monotone decreasing function of time and constitutes a suitable basis for the definition of the entropy $H = -I$. Note that this definition leads to the previous result for doubly-stochastic processes, since the uniform measure, $a_i = 1$ (all i), is obviously stationary in this case.

One can furthermore drop the requirement that the stochastic process be stationary, and even allow that there are completely different sets of states, $\{S_i^n\}$, at each time n, so that the process is now given by a sequence of matrices $T_{ij}^n$ representing the transition probability at time n from state $S_i^n$ to state $S_j^{n+1}$. In this case probability distributions change according to:

(6.3) $$P_j^{n+1} = \sum_i P_i^n T_{ij}^n .$$

If we then choose *any* time-dependent information measure which satisfies the relations:

(6.4) $$a_j^{n+1} = \sum_i a_i^n T_{ij}^n \quad (\text{all } j, n) ,$$

then the information of a probability distribution is again monotone decreasing with time. (Proof in Appendix I.)

All of these results are easily extended to the continuous case, and we see that the concept of relative information allows us to define entropy for quite general stochastic processes.

§7. *Example: Conservation of information in classical mechanics*

As a second illustrative example we consider briefly the classical mechanics of a group of particles. The system at any instant is represented by a point, $(x^1,y^1,z^1,p_x^1,p_y^1,p_z^1,\ldots,x^n,y^n,z^n,p_x^n,p_y^n,p_z^n)$, in the phase space of all position and momentum coordinates. The natural motion of the system then carries each point into another, defining a continuous transformation of the phase space into itself. According to Liouville's theorem the measure of a set of points of the phase space is invariant under this transformation.[8] This invariance of measure implies that if we begin with a probability distribution over the phase space, rather than a single point, the total information

(7.1) $$I_{\text{total}} = I_{X^1Y^1Z^1P_x^1P_y^1P_z^1 \ldots X^nY^nZ^nP_x^nP_y^nP_z^n} ,$$

which is the information of the *joint* distribution for all positions and momenta, remains *constant in time*.

8 See Khinchin [16], p. 15.

In order to see that the total information is conserved, consider any partition $\mathcal{P}$ of the phase space at one time, $t_0$, with its information relative to the phase space measure, $I^{\mathcal{P}}(t_0)$. At a later time $t_1$ a partition $\mathcal{P}'$, into the image sets of $\mathcal{P}$ under the mapping of the space into itself, is induced, for which the probabilities for the sets of $\mathcal{P}'$ are the same as those of the corresponding sets of $\mathcal{P}$, and furthermore for which the measures are the same, by Liouville's theorem. Thus corresponding to each partition $\mathcal{P}$ at time $t_0$ with information $I^{\mathcal{P}}(t_0)$, there is a partition $\mathcal{P}'$ at time $t_1$ with information $I^{\mathcal{P}'}(t_1)$, which is the same:

$$I^{\mathcal{P}'}(t_1) = I^{\mathcal{P}}(t_0) \ . \tag{7.2}$$

Due to the correspondence of the $\mathcal{P}$'s and $\mathcal{P}'$'s the supremums of each over all partitions must be equal, and by (5.5) we have proved that

$$I_{total}(t_1) = I_{total}(t_0) \ , \tag{7.3}$$

and the total information is conserved.

Now it is known that the individual (marginal) position and momentum distributions tend to decay, except for rare fluctuations, into the uniform and Maxwellian distributions respectively, for which the classical entropy is a maximum. This entropy is, however, except for the factor of Boltzman's constant, simply the negative of the marginal information

$$I_{marginal} = I_{X_1} + I_{Y_1} + I_{Z_1} + \dots + I_{P^n_x} + I_{P^n_y} + I_{P^n_z} \ , \tag{7.4}$$

which thus tends towards a minimum. But this decay of marginal information is exactly compensated by an increase of the total correlation information

$$\{total\} = I_{total} - I_{marginal} \ , \tag{7.5}$$

since the total information remains constant. Therefore, if one were to define the *total entropy* to be the negative of the total information, one could replace the usual second law of thermodynamics by a law of *conservation of total entropy*, where the increase in the standard (marginal) entropy is exactly compensated by a (negative) *correlation entropy*. The usual second law then results simply from our renunciation of all correlation knowledge (*stosszahlansatz*), and not from any intrinsic behavior of classical systems. The situation for classical mechanics is thus in sharp contrast to that of stochastic processes, which are intrinsically irreversible.

# III. QUANTUM MECHANICS

Having mathematically formulated the ideas of information and correlation for probability distributions, we turn to the field of quantum mechanics. In this chapter we assume that the states of physical systems are represented by points in a Hilbert space, and that the time dependence of the state of an isolated system is governed by a linear wave equation.

It is well known that state functions lead to distributions over eigenvalues of Hermitian operators (square amplitudes of the expansion coefficients of the state in terms of the basis consisting of eigenfunctions of the operator) which have the mathematical properties of probability distributions (non-negative and normalized). The standard interpretation of quantum mechanics regards these distributions as actually giving the probabilities that the various eigenvalues of the operator will be observed, when a measurement represented by the operator is performed.

A feature of great importance to our interpretation is the fact that a state function of a *composite* system leads to *joint* distributions over subsystem quantities, rather than independent subsystem distributions, i.e., the quantities in different subsystems may be correlated with one another. The first section of this chapter is accordingly devoted to the development of the formalism of composite systems, and the connection of composite system states and their derived joint distributions with the various possible subsystem conditional and marginal distributions. We shall see that there exist *relative state functions* which correctly give the conditional distributions for all subsystem operators, while marginal distributions can *not* generally be represented by state functions, but only by *density matrices*.

In Section 2 the concepts of information and correlation, developed in the preceding chapter, are applied to quantum mechanics, by defining information and correlation for operators on systems with prescribed states. It is also shown that for composite systems there exists a quantity which can be thought of as the fundamental correlation between subsystems, and a closely related *canonical representation* of the composite system state. In addition, a stronger form of the uncertainty principle, phrased in information language, is indicated.

The third section takes up the question of measurement in quantum mechanics, viewed as a correlation producing interaction between physical systems. A simple example of such a measurement is given and discussed. Finally some general consequences of the superposition principle are considered.

It is convenient at this point to introduce some notational conventions. We shall be concerned with points $\psi$ in a Hilbert space $\mathcal{H}$, with scalar product $(\psi_1, \psi_2)$. A *state* is a point $\psi$ for which $(\psi, \psi) = 1$. For any linear operator A we define a functional, $\langle A \rangle \psi$, called the *expectation of* A *for* $\psi$, to be:

$$\langle A \rangle \psi = (\psi, A\psi) \ .$$

A class of operators of particular interest is the class of *projection operators*. The operator $[\phi]$, called the projection on $\phi$, is defined through:

$$[\phi]\psi = (\phi, \psi)\phi \ .$$

For a complete orthonormal set $\{\phi_i\}$ and a state $\psi$ we define a *square-amplitude distribution*, $P_i$, called the distribution of $\psi$ over $\{\phi_i\}$ through:

$$P_i = |(\phi_i, \psi)|^2 = \langle [\phi_i] \rangle \psi \ .$$

In the probabilistic interpretation this distribution represents the probability distribution over the results of a measurement with eigenstates $\phi_i$, performed upon a system in the state $\psi$. (Hereafter when referring to the probabilistic interpretation we shall say briefly "the probability that the system will be found in $\phi_i$", rather than the more cumbersome phrase "the probability that the measurement of a quantity B, with eigenfunctions $\{\phi_i\}$, shall yield the eigenvalue corresponding to $\phi_i$," which is meant.)

For two Hilbert spaces $\mathcal{H}_1$ and $\mathcal{H}_2$, we form the *direct product* Hilbert space $\mathcal{H}_3 = \mathcal{H}_1 \otimes \mathcal{H}_2$ (tensor product) which is taken to be the space of all possible[1] sums of formal products of points of $\mathcal{H}_1$ and $\mathcal{H}_2$, i.e., the elements of $\mathcal{H}_3$ are those of the form $\sum_i a_i \xi_i \eta_i$ where $\xi_i \in \mathcal{H}_1$ and $\eta_i \in \mathcal{H}_2$. The scalar product in $\mathcal{H}_3$ is taken to be $\left(\sum_i a_i \xi_i \eta_i, \sum_j b_j \xi_j \eta_j\right) = \sum_{ij} a_i^* b_j (\xi_i, \xi_j)(\eta_i, \eta_j)$. It is then easily seen that if $\{\xi_i\}$ and $\{\eta_i\}$ form complete orthonormal sets in $\mathcal{H}_1$ and $\mathcal{H}_2$ respectively, then the set of all formal products $\{\xi_i \eta_j\}$ is a complete orthonormal set in $\mathcal{H}_3$. For any pair of operators A, B, in $\mathcal{H}_1$ and $\mathcal{H}_2$ there corresponds an operator $C = A \otimes B$, the direct product of A and B, in $\mathcal{H}_3$, which can be defined by its effect on the elements $\xi_i \eta_j$ of $\mathcal{H}_3$:

$$C\xi_i \eta_j = A \otimes B \xi_i \eta_j = (A\xi_i)(B\eta_j) \ .$$

§1. *Composite systems*

It is well known that if the states of a pair of systems $S_1$ and $S_2$, are represented by points in Hilbert spaces $\mathcal{H}_1$ and $\mathcal{H}_2$ respectively, then the states of the *composite system* $S = S_1 + S_2$ (the two systems $S_1$ and $S_2$ regarded as a single system S) are represented correctly by points of the direct product $\mathcal{H}_1 \otimes \mathcal{H}_2$. This fact has far reaching consequences which we wish to investigate in some detail. Thus if $\{\xi_i\}$ is a complete orthonormal set for $\mathcal{H}_1$, and $\{\eta_j\}$ for $\mathcal{H}_2$, the general state of $S = S_1 + S_2$ has the form:

$$\psi^S = \sum_{ij} a_{ij} \xi_i \eta_j \quad \left(\sum_{ij} a_{ij}^* a_{ij} = 1\right) . \tag{1.1}$$

[1] More rigorously, one considers only *finite* sums, then completes the resulting space to arrive at $\mathcal{H}_1 \otimes \mathcal{H}_2$.

In this case we shall call $P_{ij} = a_{ij}^* a_{ij}$ the *joint square-amplitude distribution* of $\psi^S$ over $\{\xi_i\}$ and $\{\eta_j\}$. In the standard probabilistic interpretation $a_{ij}^* a_{ij}$ represents the joint probability that $S_1$ will be found in the state $\xi_i$ *and* $S_2$ will be found in the state $\eta_j$. Following the probabilistic model we now derive some distributions from the state $\psi^S$. Let A be a Hermitian operator in $S_1$ with eigenfunctions $\phi_i$ and eigenvalues $\lambda_i$, and B an operator in $S_2$ with eigenfunctions $\theta_j$ and eigenvalues $\mu_j$. Then the joint distribution of $\psi^S$ over $\{\phi_i\}$ and $\{\phi_j\}$, $P_{ij}$, is:

(1.2) $$P_{ij} = P(\phi_i \text{ and } \theta_j) = |(\phi_i \theta_j, \psi^S)|^2 .$$

The *marginal* distributions, of $\psi^S$ over $\{\phi_i\}$ and of $\psi^S$ over $\{\phi_j\}$, are:

(1.3) $$P_i = P(\phi_i) = \sum_j P_{ij} = \sum_j |(\phi_i \theta_j, \psi^S)|^2 ,$$

$$P_j = P(\theta_j) = \sum_i P_{ij} = \sum_i |(\phi_i \theta_j, \psi^S)|^2 ,$$

and the *conditional distributions* $P_i^j$ and $P_j^i$ are:

(1.4) $$P_i^j = P(\phi_i \text{ conditioned on } \phi_j) = \frac{P_{ij}}{P_j} ,$$

$$P_j^i = P(\phi_j \text{ conditioned on } \phi_i) = \frac{P_{ij}}{P_i} .$$

We now define the *conditional expectation* of an operator A on $S_1$, conditioned on $\theta_j$ in $S_2$, denoted by $\mathrm{Exp}^{\theta_j}[A]$, to be:

(1.5) $$\mathrm{Exp}^{\theta_j}[A] = \sum_i \lambda_i P_i^j = (1/P_j) \sum_i P_{ij} \lambda_i$$

$$= (1/P_j) \sum_i \lambda_i |(\phi_i \theta_j, \psi^S)|^2$$

$$= (1/P_j) \sum_i |(\phi_i \theta_j, \psi^S)|^2 (\phi_i, A\phi_i) ,$$

and we define the *marginal expectation* of A on $S_1$ to be:

$$\text{(1.6)} \qquad \mathrm{Exp}\,[A] = \sum_i P_i \lambda_i = \sum_{ij} \lambda_i P_{ij} = \sum_{ij} |(\phi_i \theta_j, \psi^S)|^2 (\phi_i, A\phi_i) \; .$$

We shall now introduce projection operators to get more convenient forms of the conditional and marginal expectations, which will also exhibit more clearly the degree of dependence of these quantities upon the chosen basis $\{\phi_i \theta_j\}$. Let the operators $[\phi_i]$ and $[\phi_j]$ be the projections on $\phi_i$ in $S_1$ and $\phi_j$ in $S_2$ respectively, and let $I^1$ and $I^2$ be the identity operators in $S_1$ and $S_2$. Then, making use of the identity $\psi^S = \sum_{ij} (\phi_i \theta_j, \psi^S)\phi_i \theta_j$ for any complete orthonormal set $\{\phi_i \theta_j\}$, we have:

$$\text{(1.7)} \quad \langle[\phi_i][\theta_j]\rangle \psi^S = (\psi^S, [\phi_i][\theta_j]\psi^S) =$$

$$\left(\sum_{k\ell} (\phi_k \theta_\ell, \psi^S)\phi_k \theta_\ell, [\phi_i][\theta_j] \sum_{mn} (\phi_m \theta_n, \psi^S)\phi_m \theta_n\right)$$

$$= \sum_{k\ell mn} (\phi_k \theta_\ell, \psi^S)^* (\phi_m \theta_n, \psi^S)\delta_{km}\delta_{\ell n}\delta_{im}\delta_{jn}$$

$$= (\phi_i \theta_j, \psi^S)^* (\phi_i \theta_j, \psi^S) = P_{ij} \; ,$$

so that the joint distribution is given simply by $\langle[\phi_i][\phi_j]\rangle \psi^S$.

For the marginal distribution we have:

$$\text{(1.8)} \; P_i = \sum_j P_{ij} = \sum_j \langle[\phi_i][\theta_j]\rangle \psi^S = \langle[\phi_i]\left(\sum_j [\theta_j]\right)\rangle \psi^S = \langle[\phi_i] I^2 \rangle \psi^S \; ,$$

and we see that the marginal distribution over the $\phi_i$ is *independent* of the set $\{\theta_j\}$ chosen in $S_2$. This result has the consequence in the ordinary interpretation that the expected outcome of measurement in one subsystem of a composite system is not influenced by the choice of quantity to be measured in the other subsystem. This expectation is, in fact, the expectation for the case in which no measurement at all (identity operator) is performed in the other subsystem. Thus no measurement in $S_2$ can affect the expected outcome of a measurement in $S_1$, *so long as the result of any* $S_2$ *measurement remains unknown.* The case is quite different, however, if this result *is* known, and we must turn to the conditional distributions and expectations in such a case.

We now introduce the concept of a *relative state-function,* which will play a central role in our interpretation of pure wave mechanics. Consider a composite system $S = S_1 + S_2$ in the state $\psi^S$. To every state $\eta$ of $S_2$ we associate a state of $S_1$, $\psi^{\eta}_{rel}$, called the relative state in $S_1$ for $\eta$ in $S_2$, through:

(1.9) DEFINITION. $\psi^{\eta}_{rel} = N \sum_i (\phi_i \eta, \psi^S)\phi_i$ ,

where $\{\phi_i\}$ is *any* complete orthonormal set in $S_1$ and N is a normalization constant.[2]

The first property of $\psi^{\eta}_{rel}$ is its uniqueness,[3] i.e., its dependence upon the choice of the basis $\{\phi_i\}$ is only apparent. To prove this, choose another basis $\{\xi_k\}$, with $\phi_i = \sum_k b_{ik}\xi_k$. Then $\sum_i b^*_{ij} b_{ik} = \delta_{jk}$, and:

$$\sum_i (\phi_i \eta, \psi^S)\phi_i = \sum_i \left(\sum_j b_{ij}\xi_j \eta, \psi^S\right)\left(\sum_k b_{ik}\xi_k\right)$$

$$= \sum_{jk} \left(\sum_i b^*_{ij} b_{ik}\right)(\xi_j \eta, \psi^S)\xi_k = \sum_{jk} \delta_{jk}(\xi_j \eta, \psi^S)\xi_k$$

$$= \sum_k (\xi_k \eta, \psi^S)\xi_k \ .$$

The second property of the relative state, which justifies its name, is that $\psi^{\theta_j}_{rel}$ correctly gives the *conditional expectations* of all operators in $S_1$, conditioned by the state $\theta_j$ in $S_2$. As before let A be an operator in $S_1$ with eigenstates $\phi_i$ and eigenvalues $\lambda_i$. Then:

2 In case $\Sigma_i(\phi_i \eta, \psi^S)\phi_i = 0$ (unnormalizable) then choose any function for the relative function. This ambiguity has no consequences of any importance to us. See in this connection the remarks on p. 40.

3 Except if $\Sigma_i(\phi_i \eta, \psi^S)\phi_i = 0$. There is still, of course, no dependence upon the basis.

$$(1.10)\qquad \langle A\rangle\psi^{\theta_j}_{rel} = \left(\psi^{\theta_j}_{rel}, A\psi^{\theta_j}_{rel}\right)$$

$$= \left(N\sum_i (\phi_i\theta_j,\psi^S)\phi_i, AN\sum_{im}(\phi_m\theta_j,\psi^S)\phi_m\right)$$

$$= N^2\sum_{im}(\phi_i\theta_j,\psi^S)^*(\phi_m\theta_j,\psi^S)\lambda_m\delta_{im}$$

$$= N^2\sum_i \lambda_i P_{ij}\ .$$

At this point the normalizer $N^2$ can be conveniently evaluated by using (1.10) to compute: $\langle I^1\rangle\psi^{\theta_j}_{rel} = N^2\sum_i 1\, P_{ij} = N^2 P_j = 1$, so that

$$(1.11)\qquad N^2 = 1/P_j\ .$$

Substitution of (1.11) in (1.10) yields:

$$(1.12)\qquad \langle A\rangle\psi^{\theta_j}_{rel} = (1/P_j)\sum_i \lambda_i P_{ij} = \sum_i \lambda_i P^j_i = \mathrm{Exp}^{\theta_j}[A]\ ,$$

and we see that the conditional expectations of operators are given by the relative states. (This includes, of course, the conditional distributions themselves, since they may be obtained as expectations of projection operators.)

An important representation of a composite system state $\psi^S$, in terms of an orthonormal set $\{\theta_j\}$ in one subsystem $S_2$ and the set of relative states $\left\{\psi^{\theta_j}_{rel}\right\}$ in $S_1$ is:

$$(1.13)\qquad \psi^S = \sum_{ij}(\phi_i\theta_j,\psi^S)\phi_i\theta_j = \sum_j\left(\sum_i(\phi_i\theta_j,\psi^S)\phi_i\right)\theta_j$$

$$= \sum_j \frac{1}{N_j}\left[N_j\sum_i(\phi_i\theta_j,\psi^S)\phi_i\right]\theta_j$$

$$= \sum_j \frac{1}{N_j}\psi^{\theta_j}_{rel}\,\theta_j\ , \quad \text{where } 1/N_j^2 = P_j = \langle I^1[\theta_j]\rangle\psi^S\ .$$

Thus, for *any* orthonormal set in one subsystem, the state of the composite system is a single superposition of elements consisting of a state of the given set and its relative state in the other subsystem. (The relative states, however, are not necessarily orthogonal.) We notice further that a particular element, $\psi_{rel}^{\theta_j}\theta_j$, is quite independent of the choice of basis $\{\theta_k\}$, $k \neq j$, for the orthogonal space of $\theta_j$, since $\psi_{rel}^{\theta_j}$ depends *only* on $\theta_j$ and not on the other $\theta_k$ for $k \neq j$. We remark at this point that the ambiguity in the relative state which arises when $\sum_i (\phi_i \theta_j, \psi^S)\phi_i = 0$ (see p. 38) is unimportant for this representation, since although *any* state $\psi_{rel}^{\theta_j}$ can be regarded as the relative state in this case, the term $\psi_{rel}^{\theta_j}\theta_j$ will occur in (1.13) with coefficient zero.

Now that we have found subsystem states which correctly give conditional expectations, we might inquire whether there exist subsystem states which give marginal expectations. The answer is, unfortunately, no. Let us compute the marginal expectation of $A$ in $S_1$ using the representation (1.13):

$$\text{(1.14)} \qquad \text{Exp}\,[A] = \langle A\, I^2 \rangle \psi^S = \left( \sum_j \frac{1}{N_j} \psi_{rel}^{\theta_j} \theta_j, A\, I^2 \sum_k \frac{1}{N_k} \psi_{rel}^{\theta_k} \theta_k \right)$$

$$= \sum_{jk} \frac{1}{N_j N_k} \left( \psi_{rel}^{\theta_j}, A \psi_{rel}^{\theta_j} \right) \delta_{jk}$$

$$= \sum_j \frac{1}{N_j^2} \left( \psi_{rel}^{\theta_j}, A \psi_{rel}^{\theta_j} \right) = \sum_j P_j \langle A \rangle \psi_{rel}^{\theta_j} .$$

Now suppose that there exists a state in $S_1$, $\psi'$, which correctly gives the marginal expectation (1.14) for *all* operators $A$ (i.e., such that $\text{Exp}\,[A] = \langle A \rangle \psi'$ for all $A$). One such operator is $[\psi']$, the projection on $\psi'$, for which $\langle [\psi'] \rangle \psi' = 1$. But, from (1.14) we have that $\text{Exp}\,[\psi'] = \sum_j P_j \langle \psi' \rangle \psi_{rel}^{\theta_j}$, which is $< 1$ unless, for all $j$, $P_j = 0$ or $\psi_{rel}^{\theta_j} = \psi'$, a condition which is not generally true. Therefore *there exists in general no state for* $S_1$ *which correctly gives the marginal expectations for all operators in* $S_1$.

However, even though there is generally no single state describing marginal expectations, we see that there is always a *mixture* of states, namely the states $\psi_{rel}^{\theta_j}$ *weighted* with $P_j$, which does yield the correct expectations. The distinction between a mixture, M, of states $\phi_i$, weighted by $P_i$, and a *pure state* $\psi$ which is a superposition, $\psi = \sum a_i \phi_i$, is that there are *no interference phenomena* between the various states of a mixture. The expectation of an operator A for the mixture is $\mathrm{Exp}^M[A] = \sum_i P_i \langle A \rangle \phi_i = \sum_i P_i(\phi_i, A\phi_i)$, while the expectation for the pure state $\psi$ is $\langle A \rangle \psi = \left(\sum_i a_i\phi_i, A \sum_j a_j\phi_j\right) = \sum_{ij} a_i^* a_j(\phi_i, A\phi_j)$, which is *not* the same as that of the mixture with weights $P_i = a_i^* a_i$, due to the presence of the interference terms $(\phi_i, A\phi_j)$ for $j \neq i$.

It is convenient to represent such a mixture by a *density matrix*,[4] $\rho$. If the mixture consists of the states $\psi_j$ weighted by $P_j$, and if we are working in a basis consisting of the complete orthonormal set $\{\phi_i\}$, where $\psi_j = \sum_i a_i^j \phi_i$, then we define the elements of the density matrix for the mixture to be:

$$\rho_{k\ell} = \sum_j P_j\, a_\ell^{j*} a_k^j \qquad (a_i^j = (\phi_i, \psi_j)) \; . \tag{1.15}$$

Then if A is any operator, with matrix representation $A_{i\ell} = (\phi_i, A\phi_\ell)$ in the chosen basis, its expectation for the mixture is:

$$\begin{aligned} \mathrm{Exp}^M[A] &= \sum_j P_j(\psi_j, A\psi_j) = \sum_j P_j \left[\sum_{i\ell} a_i^{j*} a_\ell^j (\phi_i, A\phi_\ell)\right] \\ &= \sum_{i\ell}\left(\sum_j P_j\, a_i^{j*} a_\ell^j\right)(\phi_i, A\phi_\ell) = \sum_{i,\ell} \rho_{\ell i} A_{i\ell} \\ &= \mathrm{Trace}\,(\rho A) \; . \end{aligned} \tag{1.16}$$

---

[4] Also called a *statistical operator* (von Neumann [17]).

Therefore any mixture is adequately represented by a density matrix.[5] Note also that $\rho^*_{k\ell} = \rho_{\ell k}$, so that $\rho$ is Hermitian.

Let us now find the density matrices $\rho^1$ and $\rho^2$ for the subsystems $S_1$ and $S_2$ of a system $S = S_1 + S_2$ in the state $\psi^S$. Furthermore, let us choose the orthonormal bases $\{\xi_i\}$ and $\{\eta_j\}$ in $S_1$ and $S_2$ respectively, and let A be an operator in $S_1$, B an operator in $S_2$. Then:

$$\text{(1.17)}\quad \text{Exp}[A] = \langle AI^2 \rangle \psi^S = \left( \sum_{ij} (\xi_i \eta_j, \psi^S) \xi_i \eta_j, AI \sum_{\ell m} (\xi_\ell \eta_m, \psi^S) \xi_\ell \eta_m \right)$$

$$= \sum_{ij\ell m} (\xi_i \eta_j, \psi^S)^* (\xi_\ell \eta_m, \psi^S) (\xi_i, A\xi_\ell)(\eta_j, \eta_m)$$

$$= \sum_{i\ell} \left[ \sum_j (\xi_i \eta_j, \psi^S)^* (\xi_\ell \eta_j, \psi^S) \right] (\xi_i, A\xi_\ell)$$

$$= \text{Trace}\,(\rho^1 A) \ ,$$

where we have defined $\rho^1$ in the $\{\xi_i\}$ basis to be:

$$\text{(1.18)}\qquad \rho^1_{\ell i} = \sum_j (\xi_i \eta_j, \psi^S)^* (\xi_\ell \eta_j, \psi^S) \ .$$

In a similar fashion we find that $\rho^2$ is given, in the $\{\eta_j\}$ basis, by:

$$\text{(1.19)}\qquad \rho^2_{mn} = \sum_i (\xi_i \eta_n, \psi^S)^* (\xi_i \eta_m, \psi^S) \ .$$

It can be easily shown that here again the dependence of $\rho^1$ upon the choice of basis $\{\eta_j\}$ in $S_2$, and of $\rho^2$ upon $\{\xi_i\}$, is only apparent.

---

5 A better, coordinate free representation of a mixture is in terms of the operator which the density matrix represents. For a mixture of states $\psi_n$ (not necessarily orthogonal) with weights $p_n$, the density operator is $\rho = \sum_n p_n [\psi_n]$, where $[\psi_n]$ stands for the projection operator on $\psi_n$.

In summary, we have seen in this section that a state of a composite system leads to *joint* distributions over subsystem quantities which are generally not independent. Conditional distributions and expectations for subsystems are obtained from *relative states*, and subsystem marginal distributions and expectations are given by *density matrices*.

There does not, in general, exist anything like a single state for one subsystem of a composite system. That is, subsystems do *not* possess states independent of the states of the remainder of the system, so that the subsystem states are generally *correlated*. One can arbitrarily choose a state for one subsystem, and be led to the *relative state* for the other subsystem. Thus we are faced with a fundamental *relativity of states*, which is implied by the formalism of composite systems. It is meaningless to ask the absolute state of a subsystem – one can only ask the state relative to a given state of the remainder of the system.

§2. *Information and correlation in quantum mechanics*

We wish to be able to discuss information and correlation for Hermitian operators A, B,..., with respect to a state function $\psi$. These quantities are to be computed, through the formulas of the preceding chapter, from the square amplitudes of the coefficients of the expansion of $\psi$ in terms of the eigenstates of the operators.

We have already seen (p. 34) that a state $\psi$ and an orthonormal basis $\{\phi_i\}$ leads to a square amplitude distribution of $\psi$ over the set $\{\phi_i\}$:

$$P_i = |(\phi_i,\psi)|^2 = \langle[\phi_i]\rangle\psi \ , \tag{2.1}$$

so that we can define the *information of the basis* $\{\phi_i\}$ *for the state* $\psi$, $I_{\{\phi_i\}}(\psi)$, to be simply the information of this distribution relative to the uniform measure:

$$I_{\{\phi_i\}}(\psi) = \sum_i P_i \ln P_i = \sum_i |(\phi_i,\psi)|^2 \ln |(\phi_i,\psi)|^2 \ . \tag{2.2}$$

We define the *information of an operator* A, for the state $\psi$, $I_A(\psi)$, to be the information in the square amplitude distribution over its *eigenvalues*, i.e., the information of the probability distribution over the results of a determination of A which is prescribed in the probabilistic interpretation. For a *non-degenerate* operator A this distribution is the same as the distribution (2.1) over the eigenstates. But because the information is dependent only on the distribution, and not on numerical values, the information of the distribution over eigenvalues of A is precisely the information of the eigenbasis of A, $\{\phi_i\}$. Therefore:

$$(2.3)\quad I_A(\psi) = I_{\{\phi_i\}}(\psi) = \sum_i <[\phi_i]>\psi \ln <[\phi_i]>\psi \quad \text{(A non-degenerate)} .$$

We see that for fixed $\psi$, the information of all non-degenerate operators having the same set of eigenstates is the same.

In the case of *degenerate* operators it will be convenient to take, as the definition of information, the information of the square amplitude distribution over the eigenvalues *relative* to the information measure which consists of the *multiplicity* of the eigenvalues, rather than the uniform measure. This definition preserves the choice of uniform measure over the *eigenstates*, in distinction to the eigenvalues. If $\phi_{ij}$ (j from 1 to $m_i$) are a complete orthonormal set of eigenstates for A′, with distinct eigenvalues $\lambda_i$ (degenerate with respect to j), then the multiplicity of the $i^{th}$ eigenvalue is $m_i$ and the information $I_{A'}(\psi)$ is defined to be:

$$(2.4)\qquad I_{A'}(\psi) = \sum_i \left( \sum_j <[\phi_{ij}]>\psi \right) \ln \frac{\sum_j <[\phi_{ij}]>\psi}{m_i} .$$

The usefulness of this definition lies in the fact that any operator A″ which distinguishes further between any of the degenerate states of A′ leads to a refinement of the relative density, in the sense of Theorem 4, and consequently has equal or greater information. A non-degenerate operator thus represents the maximal refinement and possesses maximal information.

It is convenient to introduce a new notation for the projection operators which are *relevant* for a specified operator. As before let A have eigenfunctions $\phi_{ij}$ and distinct eigenvalues $\lambda_i$. Then define the projections $A_i$, the projections on the *eigenspaces* of different eigenvalues of A, to be:

(2.5) $$A_i = \sum_{j=1}^{m_i} [\phi_{ij}] \ .$$

To each such projection there is associated a number $m_i$, the multiplicity of the degeneracy, which is the dimension of the $i^{th}$ eigenspace. In this notation the distribution over the eigenvalues of A for the state $\psi$, $P_i$, becomes simply:

(2.6) $$P_i = P(\lambda_i) = \langle A_i \rangle \psi \ ,$$

and the information, given by (2.4), becomes:

(2.7) $$I_A = \sum_i \langle A_i \rangle \psi \ln \frac{\langle A_i \rangle \psi}{m_i} \ .$$

Similarly, for a pair of operators, A in $S_1$ and B in $S_2$, for the composite system $S = S_1 + S_2$ with state $\psi^S$, the *joint* distribution over eigenvalues is:

(2.8) $$P_{ij} = P(\lambda_i, \mu_j) = \langle A_i B_j \rangle \psi^S \ ,$$

and the marginal distributions are:

(2.9) $$P_i = \sum_j P_{ij} = \langle A_i \Big(\sum_j B_j\Big) \rangle \psi^S = \langle A_i I^2 \rangle \psi^S \ ,$$

$$P_j = \sum_i P_{ij} = \langle \Big(\sum_i A_i\Big) B_j \rangle \psi^S = \langle I^1 B_j \rangle \psi^S \ .$$

The *joint* information, $I_{AB}$, is given by:

(2.10) $$I_{AB} = \sum_{ij} P_{ij} \ln \frac{P_{ij}}{m_i n_j} = \sum_{ij} \langle A_i B_j \rangle \psi^S \ln \frac{\langle A_i B_j \rangle \psi^S}{m_i n_j} \ ,$$

where $m_i$ and $n_j$ are the multiplicities of the eigenvalues $\lambda_i$ and $\mu_j$. The marginal information quantities are given by:

$$I_A = \sum_i \langle A_i I^2 \rangle \psi^S \ln \frac{\langle A_i I^2 \rangle \psi^S}{m_i} , \tag{2.11}$$

$$I_B = \sum_j \langle I^1 B_j \rangle \psi^S \ln \frac{\langle I^1 B_j \rangle \psi^S}{n_j} ,$$

and finally the correlation, $\{A, B\}\psi^S$ is given by:

$$\{A,B\}\psi^S = \sum_{ij} P_{ij} \ln \frac{P_{ij}}{P_i P_j} = \sum_{ij} \langle A_i B_j \rangle \psi^S \ln \frac{\langle A_i B_j \rangle \psi^S}{\langle A_i I \rangle \psi^S \langle I B_j \rangle \psi^S} , \tag{2.12}$$

where we note that the expression does not involve the multiplicities, as do the information expressions, a circumstance which simply reflects the independence of correlation on any information measure. These expressions of course generalize trivially to distributions over more than two variables (composite systems of more than two subsystems).

In addition to the correlation of pairs of subsystem operators, given by (2.12), there always exists a unique quantity $\{S_1, S_2\}$, the *canonical correlation*, which has some special properties and may be regarded as the fundamental correlation between the two subsystems $S_1$ and $S_2$ of the composite system S. As we remarked earlier a density matrix is Hermitian, so that there is a representation in which it is diagonal.[6] In

---

[6] The density matrix of a subsystem always has a pure discrete spectrum, if the composite system is in a state. To see this we note that the choice of any orthonormal basis in $S_2$ leads to a discrete (i.e., denumerable) set of relative states in $S_1$. The density matrix in $S_1$ then represents *this* discrete mixture, $\psi^{\theta_j}_{rel}$ weighted by $P_j$. This means that the expectation of the identity, $\mathrm{Exp}[I] = \Sigma_j P_j(\psi^{\theta_j}_{rel}, I\psi^{\theta_j}_{rel}) = \Sigma_j P_j = 1 = \mathrm{Trace}(\rho I) = \mathrm{Trace}(\rho)$. Therefore $\rho$ has a finite trace and is a completely continuous operator, having necessarily a pure discrete spectrum. (See von Neumann [17], p. 89, footnote 115.)

particular, for the decomposition of S (with state $\psi^S$) into $S_1$ and $S_2$, we can choose a representation in which both $\rho^{S_1}$ and $\rho^{S_2}$ are diagonal. (This choice is always possible because $\rho^{S_1}$ is independent of the basis in $S_2$ and vice-versa.) Such a representation will be called a *canonical representation.* This means that it is always possible to represent the state $\psi^S$ by a *single* superposition:

$$\psi^S = \sum_i a_i \xi_i \eta_i \;, \tag{2.13}$$

where *both* the $\{\xi_i\}$ and the $\{\eta_i\}$ constitute orthonormal sets of states for $S_1$ and $S_2$ respectively.

To construct such a representation choose the basis $\{\eta_i\}$ for $S_2$ so that $\rho^{S_2}$ is diagonal:

$$\rho_{ij}^{S_2} = \lambda_i \delta_{ij} \;, \tag{2.14}$$

and let the $\xi_i$ be the *relative* states in $S_1$ for the $\eta_i$ in $S_2$:

$$\xi_i = N_i \sum_j (\phi_j \eta_i, \psi^S)\phi_j \quad \text{(any basis } \{\phi_j\}) \;. \tag{2.15}$$

Then, according to (1.13), $\psi^S$ is represented in the form (2.13) where the $\{\eta_i\}$ are orthonormal by choice, and the $\{\xi_i\}$ are normal since they are relative states. We therefore need only show that the states $\{\xi_i\}$ are orthogonal:

$$\begin{aligned}(\xi_j, \xi_k) &= \left(N_j \sum_\ell (\phi_\ell \eta_j, \psi^S)\phi_\ell, N_k \sum_m (\phi_m \eta_k, \psi^S)\phi_m\right) \\ &= \sum_{\ell m} N_j^* N_k (\phi_\ell \eta_j, \psi^S)^* (\phi_m \eta_k, \psi^S)\delta_{\ell m} \\ &= N_j^* N_k \sum_\ell (\phi_\ell \eta_j, \psi^S)^* (\phi_\ell \eta_k, \psi^S) \\ &= N_j^* N_k \rho_{kj}^{S_2} = N_j^* N_k \lambda_k \delta_{kj} = 0, \text{ for } j \neq k \;,\end{aligned} \tag{2.16}$$

since we supposed $\rho^{S_2}$ to be diagonal in this representation. We have therefore constructed a canonical representation (2.13).

The density matrix $\rho^{S_1}$ is also automatically diagonal, by the choice of representation consisting of the basis in $S_2$ which makes $\rho^{S_2}$ diagonal and the corresponding relative states in $S_1$. Since $\{\xi_i\}$ are orthonormal we have:

$$(2.17) \qquad \rho^{S_1}_{ij} = \sum_k (\xi_i \eta_k, \psi^S)^* (\xi_j \eta_k, \psi^S) =$$

$$\sum_k \left(\xi_i \eta_k, \sum_m a_m \xi_m \eta_m\right)^* \left(\xi_j \eta_k, \sum_\ell a_\ell \xi_\ell \eta_\ell\right)$$

$$= \sum_{k\ell m} a_m^* a_\ell \delta_{im} \delta_{km} \delta_{j\ell} \delta_{k\ell} = \sum_k a_i^* a_j \delta_{ki} \delta_{kj}$$

$$= a_i^* a_i \delta_{ij} = P_i \delta_{ij} ,$$

where $P_i = a_i^* a_i$ is the marginal distribution over the $\{\xi_i\}$. Similar computation shows that the elements of $\rho^{S_2}$ are the *same*:

$$(2.18) \qquad \rho^{S_2}_{k\ell} = a_k^* a_k \delta_{k\ell} = P_k \delta_{k\ell} .$$

Thus in the canonical representation both density matrices are diagonal and have the same elements, $P_k$, which give the marginal square amplitude distribution over both of the sets $\{\xi_i\}$ and $\{\eta_i\}$ forming the basis of the representation.

Now, any pair of operators, $\widetilde{A}$ in $S_1$ and $\widetilde{B}$ in $S_2$, which have as non-degenerate eigenfunctions the sets $\{\xi_i\}$ and $\{\eta_j\}$ (i.e., operators which define the canonical representation), are "perfectly" correlated in the sense that there is a one-one correspondence between their eigenvalues. The joint square amplitude distribution for eigenvalues $\lambda_i$ of $\widetilde{A}$ and $\mu_j$ of $\widetilde{B}$ is:

$$(2.19) \qquad P(\lambda_i \text{ and } \mu_j) = P(\xi_i \text{ and } \eta_j) = P_{ij} = a_i^* a_i \delta_{ij} = P_i \delta_{ij} .$$

Therefore, the correlation between these operators, $\{\tilde{A},\tilde{B}\}\psi^S$ is:

$$(2.20)\quad \{\tilde{A},\tilde{B}\}\psi^S = \sum_{ij} P(\lambda_i \text{ and } \mu_j) \ln \frac{P(\lambda_i \& \mu_j)}{P(\lambda_i)P(\mu_j)} = \sum_{ij} P_i \delta_{ij} \ln \frac{P_i \delta_{ij}}{P_i P_j}$$

$$= -\sum_i P_i \ln P_i .$$

We shall denote this quantity by $\{S_1, S_2\}\psi^S$ and call it the *canonical correlation* of the subsystems $S_1$ and $S_2$ for the system state $\psi^S$. It is the correlation between any pair of non-degenerate subsystem operators which define the canonical representation.

In the canonical representation, where the density matrices are diagonal ((2.17) and (2.18)), the canonical correlation is given by:

$$(2.21)\quad \{S_1,S_2\}\psi^S = -\sum_i P_i \ln P_i = -\text{Trace}(\rho^{S_1} \ln \rho^{S_1})$$

$$= -\text{Trace}(\rho^{S_2} \ln \rho^{S_2}) .$$

But the trace is invariant for unitary transformations, so that (2.21) holds independently of the representation, and we have therefore established the *uniqueness* of $\{S_1,S_2\}\psi^S$.

It is also interesting to note that the quantity $-\text{Trace}(\rho \ln \rho)$ is (apart from a factor of Boltzman's constant) just the *entropy* of a mixture of states characterized by the density matrix $\rho$.[7] Therefore the entropy of the mixture characteristic of a subsystem $S_1$ for the state $\psi^S = \psi^{S_1+S_2}$ is exactly matched by a correlation information $\{S_1,S_2\}$, which represents the correlation between any pair of operators $\tilde{A}$, $\tilde{B}$, which define the canonical representation. The situation is thus quite similar to that of classical mechanics.[8]

7 See von Neumann [17], p. 296.

8 Cf. Chapter II, §7.

Another special property of the canonical representation is that any operators $\widetilde{A}$, $\widetilde{B}$ defining a canonical representation have *maximum marginal information,* in the sense that for any other discrete spectrum operators, A on $S_1$, B on $S_2$, $I_A \leqq I_{\widetilde{A}}$ and $I_B \leqq I_{\widetilde{B}}$. If the canonical representation is (2.13), with $\{\xi_i\}$, $\{\eta_i\}$ non-degenerate eigenfunctions of $\widetilde{A}$, $\widetilde{B}$, respectively, and A, B any pair of non-degenerate operators with eigenfunctions $\{\phi_k\}$ and $\{\theta_\ell\}$, where $\xi_i = \sum_k c_{ik}\phi_k$, $\eta_i = \sum_\ell d_{i\ell}\theta_\ell$, then $\psi^S$ in $\phi, \theta$ representation is:

$$\psi^S = \sum_{ik\ell} a_i c_{ik} d_{i\ell} \phi_k \theta_\ell = \sum_{k\ell}\left(\sum_i a_i c_{ik} d_{i\ell}\right)\phi_k\theta_\ell , \tag{2.22}$$

and the joint square amplitude distribution for $\phi_k, \theta_\ell$ is:

$$P_{k\ell} = \left|\left(\sum_i a_i c_{ik} d_{i\ell}\right)\right|^2 = \sum_{im} a_i^* a_m c_{ik}^* c_{mk} d_{i\ell}^* d_{m\ell} , \tag{2.23}$$

while the marginals are:

$$P_k = \sum_\ell P_{k\ell} = \sum_{im} a_i^* a_m c_{ik}^* c_{mk} \sum_\ell d_{i\ell}^* d_{m\ell} \tag{2.24}$$

$$= \sum_{im} a_i^* a_m c_{ik}^* c_{mk} \delta_{im} = \sum_i a_i^* a_i c_{ik}^* c_{ik} ,$$

and similarly

$$P_\ell = \sum_k P_{k\ell} = \sum_i a_i^* a_i d_{i\ell}^* d_{i\ell} . \tag{2.25}$$

Then the marginal information $I_A$ is:

$$I_A = \sum_k P_k \ln P_k = \sum_k \left(\sum_i a_i^* a_i c_{ik}^* c_{ik}\right) \ln \left(\sum_i a_i^* a_i c_{ik}^* c_{ik}\right) \tag{2.26}$$

$$= \sum_k \left(\sum_i a_i^* a_i T_{ik}\right) \ln \left(\sum_i a_i^* a_i T_{ik}\right) ,$$

where $T_{ik} = c_{ik}^* c_{ik}$ is doubly-stochastic $(\sum_i T_{ik} = \sum_k T_{ik} = 1$ follows from unitary nature of the $c_{ik}$). Therefore (by Corollary 2, §4, Appendix I):

(2.27)
$$I_A = \sum_k \left(\sum_i a_i^* a_i T_{ik}\right) \ln \left(\sum_i a_i^* a_i T_{ik}\right)$$
$$\leq \sum_i a_i^* a_i \ln a_i^* a_i = I_{\tilde{A}} ,$$

and we have proved that $\tilde{A}$ has maximal marginal information among the discrete spectrum operators. Identical proof holds for $\tilde{B}$.

While this result was proved only for non-degenerate operators, it is immediately extended to the degenerate case, since as a consequence of our definition of information for a degenerate operator, (2.4), its information is still less than that of an operator which removes the degeneracy. We have thus proved:

THEOREM. $I_A \leq I_{\tilde{A}}$, *where* $\tilde{A}$ *is any non-degenerate operator defining the canonical representation, and* A *is any operator with discrete spectrum.*

We conclude the discussion of the canonical representation by conjecturing that in addition to the maximum marginal information properties of $\tilde{A}$, $\tilde{B}$, which define the representation, they are also *maximally correlated*, by which we mean that for any pair of operators C in $S_1$, D in $S_2$, $\{C,D\} \leq \{\tilde{A},\tilde{B}\}$, i.e.,:

(2.28) CONJECTURE.[9] $\{C,D\}\psi^S \leq \{\tilde{A},\tilde{B}\}\psi^S = \{S_1,S_2\}\psi^S$ for *all* C on $S_1$, D on $S_2$.

As a final topic for this section we point out that the uncertainty principle can probably be phrased in a stronger form in terms of information. The usual form of this principle is stated in terms of *variances*, namely:

[9] The relations $\{C,\tilde{B}\} \leq \{\tilde{A},\tilde{B}\} = \{S_1,S_2\}$ and $\{\tilde{A},D\} \leq \{S_1,S_2\}$ for all C on $S_1$, D on $S_2$, can be proved easily in a manner analogous to (2.27). These do not, however, necessarily imply the general relation (2.28).

(2.29) $$\sigma_x^2 \sigma_k^2 \geq \frac{1}{4} \qquad \text{for all } \psi(x) ,$$

$$\text{where} \quad \sigma_x^2 = \langle x^2 \rangle \psi - [\langle x \rangle \psi]^2 \quad \text{and}$$

$$\sigma_k^2 = \left\langle \left(-i \frac{\partial}{\partial x}\right)^2 \right\rangle \psi - \left[\left\langle -i \frac{\partial}{\partial x} \right\rangle \psi\right]^2 = \left\langle \left(\frac{P}{\hbar}\right)^2 \right\rangle \psi - \left[\left\langle \frac{P}{\hbar} \right\rangle \psi\right]^2 .$$

The conjectured information form of this principle is:

(2.30) $$I_x + I_k \leqq \ln (1/\pi e) \qquad \text{for all } \psi(x).$$

Although this inequality has not yet been proved with complete rigor, it is made highly probable by the circumstance that *equality* holds for $\psi(x)$ of the form $\psi(x) = (1/2\pi)^{\frac{1}{4}}$ exponent $\left[\frac{x^2}{4\sigma_x^2}\right]$ the so called "minimum uncertainty packets" which give normal distributions for both position and momentum, and that furthermore the first variation of $(I_x + I_k)$ vanishes for such $\psi(x)$. (See Appendix I, §6.) Thus, although $\ln (1/\pi e)$ has not been proved an absolute maximum of $I_x + I_k$, it is at least a stationary value.

The principle (2.30) is *stronger* than (2.29), since it implies (2.29) but is not implied by it. To see that it implies (2.29) we use the well known fact (easily established by a variation calculation) that, for fixed variance $\sigma^2$, the distribution of minimum information is a normal distribution, which has information $I = \ln(1/\sigma\sqrt{2\pi e})$. This gives us the general inequality involving information and variance:

(2.31) $$I \geqq \ln(1/\sigma\sqrt{2\pi e}) \qquad \text{(for all distributions)} .$$

Substitution of (2.31) into (2.30) then yields:

(2.32) $$\ln(1/\sigma_x\sqrt{2\pi e}) + \ln(1/\sigma_k\sqrt{2\pi e}) \leqq I_x + I_k \leqq \ln(1/\pi e)$$

$$\Rightarrow (1/\sigma_x\sigma_k 2\pi e) \leqq (1/\pi e) \Rightarrow \sigma_x^2\sigma_k^2 \geqq \frac{1}{4} ,$$

so that our principle implies the standard principle (2.29).

To show that (2.29) does *not imply* (2.30) it suffices to give a counterexample. The distributions $P(x) = \frac{1}{2}\delta(x) + \frac{1}{2}\delta(x-10)$ and $P(k) = \frac{1}{2}\delta(k) + \frac{1}{2}\delta(k-10)$, which consist simply of spikes at 0 and 10, clearly satisfy (2.29), while they both have infinite information and thus do *not* satisfy (2.30). Therefore it is possible to have arbitrarily high information about *both* x and k (or p) and *still satisfy* (2.13). We have, then, another illustration that information concepts are more powerful and more natural than the older measures based upon variance.

## §3. *Measurement*

We now consider the question of measurement in quantum mechanics, which we desire to treat as a natural process within the theory of pure wave mechanics. From our point of view there is no fundamental distinction between "measuring apparata" and other physical systems. For us, therefore, a measurement is simply a special case of interaction between physical systems – an interaction which has the property of *correlating* a quantity in one subsystem with a quantity in another.

Nearly every interaction between systems produces *some* correlation however. Suppose that at some instant a pair of systems are independent, so that the composite system state function is a product of subsystem states ($\psi^S = \psi^{S_1}\psi^{S_2}$). Then this condition obviously holds only instantaneously if the systems are interacting[10] – the independence is immediately destroyed and the systems become correlated. We could, then, take the position that the two interacting systems are continually "measuring" one another, if we wished. At each instant t we could put the composite system into canonical representation, and choose a pair of operators $\widetilde{A}(t)$

---

[10] If $U_t^S$ is the unitary operator generating the time dependence for the state function of the composite system $S = S_1 + S_2$, so that $\psi_t^S = U_t^S \psi_0^S$, then we shall say that $S_1$ and $S_2$ have not interacted during the time interval $[0,t]$ if and only if $U_t^S$ is the direct product of two subsystem unitary operators, i.e., if $U_t^S = U_t^{S_1} \otimes U_t^{S_2}$.

*in* $S_1$ *and* $\tilde{B}(t)$ *in* $S_2$ *which* define this representation. We might then reasonably assert that the quantity $\tilde{A}$ in $S_1$ is measured by $\tilde{B}$ in $S_2$ (or vice-versa), since there is a one-one correspondence between their values.

Such a viewpoint, however, does not correspond closely with our intuitive idea of what constitutes "measurement," since the quantities $\tilde{A}$ and $\tilde{B}$ which turn out to be measured depend not only on the time, but also upon the initial state of the composite system. A more reasonable position is to associate the term "measurement" with a fixed interaction H between systems,[11] and to define the "measured quantities" not as those quantities $\tilde{A}(t)$, $\tilde{B}(t)$ which are instantaneously canonically correlated, but as the limit of the instantaneous canonical operators as the time goes to infinity, $\tilde{A}_\infty$, $\tilde{B}_\infty$ – provided that this limit exists and is independent of the initial state.[12] In such a case we are able to associate the "measured quantities," $\tilde{A}_\infty$, $\tilde{B}_\infty$, with the interaction H independently of the actual system states and the time. We can therefore say that H is an interaction which causes the quantity $\tilde{A}_\infty$ in $S_1$ to be measured by $\tilde{B}_\infty$ in $S_2$. For finite times of interaction the measurement is only approximate, approaching exactness as the time of interaction increases indefinitely.

There is still one more requirement that we must impose on an interaction before we shall call it a measurement. If H is to produce a measurement of A in $S_1$ by B in $S_2$, then we require that H shall

---

11 Here H means the *total* Hamiltonian of S, not just an interaction part.

12 Actually, rather than referring to canonical operators $\tilde{A}$, $\tilde{B}$, which are not unique, we should refer to the *bases* of the canonical representation, $\{\xi_i\}$ in $S_1$ and $\{\eta_j\}$ in $S_2$, since *any* operators $\tilde{A} = \Sigma_i \lambda_i [\xi_i]$, $\tilde{B} = \Sigma_j \mu_j [\eta_j]$, with the completely arbitrary eigenvalues $\lambda_i$, $\mu_j$, are canonical. The limit then refers to the limit of the canonical bases, if it exists in some appropriate sense. However, we shall, for convenience, continue to represent the canonical bases by operators.

never decrease the information in the marginal distribution of A. If H is to produce a measurement of A by correlating it with B, we expect that a knowledge of B shall give us more information about A than we had before the measurement took place, since otherwise the measurement would be useless. Now, H might produce a correlation between A and B by simply destroying the marginal information of A, without improving the expected conditional information of A given B, so that a knowledge of B would give us no more information about A than we possessed originally. Therefore in order to be sure that we will gain information about A by knowing B, when B has become correlated with A, it is necessary that the marginal information about A has not decreased. The expected information gain in this case is assured to be not less than the correlation $\{A,B\}$.

The restriction that H shall not decrease the marginal information of A has the interesting consequence that the eigenstates of A will not be distrubed, i.e., initial states of the form $\psi_0^S = \phi\eta_0$, where $\phi$ is an eigenfunction of A, must be transformed after any time interval into states of the form $\psi_t^S = \phi\eta_t$, since otherwise the marginal information of A, which was initially perfect, would be decreased. This condition, in turn, is connected with the *repeatability* of measurements, as we shall subsequently see, and could alternately have been chosen as the condition for measurement.

We shall therefore accept the following definition. An interaction H is a measurement of A in $S_1$ by B in $S_2$ if H does not destroy the marginal information of A (equivalently: if H does not disturb the eigenstates of A in the above sense) and if furthermore the correlation $\{A,B\}$ increases toward its maximum[13] with time.

---

13 The maximum of $\{A,B\}$ is $-I_A$ if A has only a discrete spectrum, and $\infty$ if it has a continuous spectrum.

We now illustrate the production of correlation with an example of a simplified measurement due to von Neumann.[14] Suppose that we have a system of only one coordinate, q, (such as position of a particle), and an apparatus of one coordinate r (for example the position of a meter needle). Further suppose that they are initially independent, so that the combined wave function is $\psi_0^{S+A} = \phi(q)\eta(r)$, where $\phi(q)$ is the initial system wave function, and $\eta(r)$ is the initial apparatus function. Finally suppose that the masses are sufficiently large or the time of interaction sufficiently small that the kinetic portion of the energy may be neglected, so that during the time of measurement the Hamiltonian shall consist only of an interaction, which we shall take to be:

(3.1) $$H_I = -i\hbar q \frac{\partial}{\partial r} .$$

Then it is easily verified that the state $\psi_t^{S+A}(q,r)$:

(3.2) $$\psi_t^{S+A}(q,r) = \phi(q)\eta(r-qt) .$$

is a solution of the Schrödinger equation

(3.3) $$i\hbar \frac{\partial \psi_t^{S+A}}{\partial t} = H_I \psi_t^{S+A}$$

for the specified initial conditions at time $t = 0$.

Translating (3.2) into square amplitudes we get:

(3.4) $$P_t(q,r) = P_1(q) P_2(r-qt) ,$$

where $$P_1(q) = \phi^*(q)\phi(q) , \quad P_2(r) = \eta^*(r)\eta(r) ,$$

and $$P_t(q,r) = \psi_t^{S+A^*}(q,r)\psi_t^{S+A}(q,r) ,$$

---

14 von Neumann [17], p. 442.

and we note that for a fixed time, t, the conditional square amplitude distribution for r has been translated by an amount depending upon the value of q, while the marginal distribution for q has been unaltered. We see thus that a correlation has been introduced between q and r by this interaction, which allows us to interpret it as a measurement. It is instructive to see quantitatively how fast this correlation takes place. We note that:

$$I_{QR}(t) = \iint P_t(q,r) \ln P_t(q,r)\, dqdr \tag{3.5}$$

$$= \iint P_1(q)P_2(r-qt) \ln P_1(q)P_2(r-qt)\, dqdr$$

$$= \iint P_1(q)P_2(\omega) \ln P_1(q)P_2(\omega)\, dqd\omega$$

$$= I_{QR}(0) \ ,$$

so that the information of the joint distribution does not change. Furthermore, since the marginal distribution for q is unchanged:

$$I_Q(t) = I_Q(0) \ , \tag{3.6}$$

and the only quantity which can change is the marginal information, $I_R$, of r, whose distribution is:

$$P_t(r) = \int P_t(r,q)\, dq = \int P_1(q)P_2(r-qt)\, dq \ . \tag{3.7}$$

Application of a special inequality (proved in §5, Appendix I) to (3.7) yields the relation:

$$I_R(t) \leqq I_Q(0) - \ln t \ , \tag{3.8}$$

so that, except for the additive constant $I_Q(0)$, the marginal information $I_R$ tends to decrease at least as fast as ln t with time during the interaction. This implies the relation for the correlation:

$$(3.9) \quad \{Q,R\}_t = I_{QR}(t) - I_Q(t) - I_R(t) \geq I_{RQ}(t) - I_Q(t) - I_Q(0) + \ln t \ .$$

But at $t = 0$ the distributions for R and Q were independent, so that $I_{RQ}(0) = I_R(0) + I_Q(0)$. Substitution of this relation, (3.5), and (3.6) into (3.9) then yields the final result:

$$(3.10) \qquad \{Q,R\}_t \geq I_R(0) - I_Q(0) + \ln t \ .$$

Therefore the correlation is built up at least as fast as $\ln t$, except for an additive constant representing the difference of the information of the initial distributions $P_2(r)$ and $P_1(q)$. Since the correlation goes to infinity with increasing time, and the marginal system distribution is not changed, the interaction (3.1) satisfies our definition of a measurement of q by r.

Even though the apparatus does not indicate any definite system value (since there are no independent system or apparatus states), one can nevertheless look upon the total wave function (3.2) as a *superposition* of pairs of subsystem states, each element of which has a definite q value and a correspondingly displaced apparatus state.[15] Thus we can write (3.2) as:

$$(3.11) \qquad \psi_t^{S+A} = \int \phi(q')\delta(q-q')\eta(r-q't)\,dq' \ ,$$

which is a superposition of states $\psi_{q'} = \delta(q-q')\eta(r-q't)$. Each of these elements, $\psi_{q'}$, of the superposition describes a state in which the system has the definite value $q = q'$, and in which the apparatus has a state that is displaced from its original state by the amount $q't$. These elements $\psi_{q'}$ are then superposed with coefficients $\phi(q')$ to form the total state (3.11).

15 See discussion of relative states, p. 38.

Conversely, if we transform to the representation where the *apparatus* is definite, we write (3.2) as:

$$\psi_t^{S+A} = \int (1/N_{r'})\xi^{r'}(q)\delta(r-r')\,dr' , \tag{3.12}$$

where
$$\xi^{r'}(q) = N_{r'}\phi(q)\eta(r'-qt)$$

and
$$(1/N_{r'})^2 = \int \phi^*(q)\phi(q)\eta^*(r'-qt)\eta(r-qt)\,dq .$$

Then the $\xi^{r'}(q)$ are the relative system state functions for the apparatus states $\delta(r-r')$ of definite value $r = r'$.

We notice that these relative system states, $\xi^{r'}(q)$, are nearly eigenstates for the values $q = r'/t$, if the degree of correlation between $q$ and $r$ is sufficiently high, i.e., if $t$ is sufficiently large, or $\eta(r)$ sufficiently sharp (near $\delta(r)$ ) then $\xi^{r'}(q)$ is nearly $\delta(q-r'/t)$.

This property, that the relative system states become approximate eigenstates of the measurement, is in fact common to all measurements. If we adopt as a measure of the nearness of a state $\psi$ to being an eigenfunction of an operator A the information $I_A(\psi)$, which is reasonable because $I_A(\psi)$ measures the sharpness of the distribution of A for $\psi$, then it is a consequence of our definition of a measurement that the relative system states tend to become eigenstates as the interaction proceeds. Since $\text{Exp}[I_Q^r] = I_Q + \{Q,R\}$, and $I_Q$ remains constant while $\{Q,R\}$ tends toward its maximum (or infinity) during the interaction, we have that $\text{Exp}[I_Q^r]$ tends to a maximum (or infinity). But $I_Q^r$ is just the information in the relative system states, which we have adopted as a measure of the nearness to an eigenstate. Therefore, at least in expectation, the relative system states approach eigenstates.

We have seen that (3.12) is a superposition of states $\psi_{r'}$, *for each of which* the apparatus has recorded a definite value $r'$, and the system is left in approximately the eigenstate of the measurement corresponding to $q = r'/t$. The discontinuous "jump" into an eigenstate is thus only a relative proposition, dependent upon our decomposition of the total wave function into the superposition, and relative to a particularly chosen apparatus value. So far as the complete theory is concerned all elements of the superposition exist simultaneously, and the entire process is quite continuous.

We have here only a special case of the following general principle which will hold for any situation which is treated entirely wave mechanically:

PRINCIPLE. For any situation in which the existence of a property $R_i$ for a subsystem $S_1$ of a composite system $S$ will imply the later property $Q_i$ for $S$, then it is also true that an initial state for $S_1$ of the form $\psi^{S_1} = \sum a_i \psi^{S_1}_{[R_i]}$ which is a *superposition* of states with the properties $R_i$, will result in a later state for $S$ of the form $\psi^S = \sum_i a_i \psi^S_{[Q_i]}$, which is *also a superposition*, of states with the property $Q_i$. That is, for any arrangement of an interaction between two systems $S_1$ and $S_2$, which has the property that each initial state $\phi_i^{S_1} \psi^{S_2}$ will result in a final situation with total state $\psi_i^{S_1+S_2}$, an initial state of $S_1$ of the form $\sum_i a_i \phi_i^{S_1}$ will lead, after interaction, to the superposition $\sum_i a_i \psi_i^{S_1+S_2}$ for the whole system.

This follows immediately from the superposition principle for solutions of a linear wave equation. It therefore holds for *any* system of quantum mechanics for which the superposition principle holds, both particle and field theories, relativistic or not, and is applicable to all physical systems, regardless of size.

This principle has the far reaching implication that for any possible measurement, for which the initial system state is not an eigenstate, the resulting state of the composite system leads to *no* definite system state nor any definite apparatus state. The system will not be put into one or another of its eigenstates with the apparatus indicating the corresponding value, and nothing resembling Process 1 can take place.

To see that this is indeed the case, suppose that we have a measuring arrangement with the following properties. The initial apparatus state is $\psi_0^A$. If the system is initially in an eigenstate of the measurement, $\phi_i^S$, then after a specified time of interaction the total state $\phi_i^S \psi_0^A$ will be transformed into a state $\phi_i^S \psi_i^A$, i.e., the system eigenstate shall not be disturbed, and the apparatus state is changed to $\psi_i^A$, which is different for each $\phi_i^S$. ($\psi_i^A$ may for example be a state describing the apparatus as indicating, by the position of a meter needle, the eigenvalue of $\phi_i^S$.) However, if the initial system state is *not an eigenstate* but a superposition $\sum_i a_i \phi_i^S$, then the final composite system state is *also a superposition*, $\sum_i a_i \phi_i^S \psi_i^\lambda$. This follows from the superposition principle since all we need do is superpose our solutions for the eigenstates, $\phi_i^S \psi_0^A \rightarrow \phi_i^S \psi_i^A$, to arrive at the solution, $\sum_i a_i \phi_i^S \psi_0^A \rightarrow \sum_i a_i \phi_i^S \psi_i^A$, for the general case. Thus in general after a measurement has been performed there will be no definite system state nor any definite apparatus state, even though there is a correlation. It seems as though nothing can ever be settled by such a measurement. Furthermore this result is independent of the *size* of the apparatus, and remains true for apparatus of quite macroscopic dimensions.

Suppose, for example, that we coupled a spin measuring device to a cannonball, so that if the spin is up the cannonball will be shifted one foot to the left, while if the spin is down it will be shifted an equal distance to the right. If we now perform a measurement with this arrangement upon a particle whose spin is a superposition of up and down, then the resulting total state will also be a superposition of two states, one in which the cannonball is to the left, and one in which it is to the right. There is no definite position for our macroscopic cannonball!

This behavior seems to be quite at variance with our observations, since macroscopic objects always appear to us to have definite positions. Can we reconcile this prediction of the purely wave mechanical theory with experience, or must we abandon it as untenable? In order to answer this question we must consider the problem of observation itself within the framework of the theory.

# IV. OBSERVATION

We shall now give an abstract treatment of the problem of observation. In keeping with the spirit of our investigation of the consequences of pure wave mechanics we have no alternative but to introduce observers, considered as purely physical systems, into the theory.

We saw in the last chapter that in general a measurement (coupling of system and apparatus) had the outcome that neither the system nor the apparatus had any definite state after the interaction – a result seemingly at variance with our experience. However, we do not do justice to the theory of pure wave mechanics until we have investigated what the theory itself says about the *appearance* of phenomena to observers, rather than hastily concluding that the theory must be incorrect because the actual states of systems as given by the theory seem to contradict our observations.

We shall see that the introduction of observers can be accomplished in a reasonable manner, and that the theory then predicts that the *appearance* of phenomena, as the subjective experience of these observers, is precisely in accordance with the predictions of the usual probabilistic interpretation of quantum mechanics.

## §1. *Formulation of the problem*

We are faced with the task of making deductions about the appearance of phenomena on a subjective level, to observers which are considered as purely physical systems and are treated within the theory. In order to accomplish this it is necessary to identify some objective properties of such an observer (states) with subjective knowledge (i.e., perceptions). Thus, in order to say that an observer O has observed the event $a$, it is necessary that the state of O has become changed from its former state to a new state which is dependent upon $a$.

It will suffice for our purposes to consider our observers to possess memories (i.e., parts of a relatively permanent nature whose states are in correspondence with the past experience of the observer). In order to make deductions about the subjective experience of an observer it is sufficient to examine the contents of the memory.

As models for observers we can, if we wish, consider automatically functioning machines, possessing sensory apparata and coupled to recording devices capable of registering past sensory data and machine configurations. We can further suppose that the machine is so constructed that its present actions shall be determined not only by its present sensory data, but by the contents of its memory as well. Such a machine will then be capable of performing a sequence of observations (measurements), and furthermore of deciding upon its future experiments on the basis of past results. We note that if we consider that current sensory data, as well as machine configuration, is immediately recorded in the memory, then the actions of the machine at a given instant can be regarded as a function of the memory contents only, and all relevant experience of the machine is contained in the memory.

For such machines we are justified in using such phrases as "the machine has perceived A" or "the machine is aware of A" if the occurrence of A is represented in the memory, since the future behavior of the machine will be based upon the occurrence of A. In fact, all of the customary language of subjective experience is quite applicable to such machines, and forms the most natural and useful mode of expression when dealing with their behavior, as is well known to individuals who work with complex automata.

When dealing quantum mechanically with a system representing an observer we shall ascribe a state function, $\psi^O$, to it. When the State $\psi^O$ describes an observer whose memory contains representations of the events A,B,...,C we shall denote this fact by appending the memory sequence in brackets as a subscript, writing:

$$\psi^O_{[A,B,\ldots,C]} \; .$$

The symbols A,B,...,C, which we shall assume to be ordered time wise, shall therefore stand for memory configurations which are in correspondence with the past experience of the observer. These configurations can be thought of as punches in a paper tape, impressions on a magnetic reel, configurations of a relay switching circuit, or even configurations of brain cells. We only require that they be capable of the interpretation "The observer has experienced the succession of events A,B,...,C." (We shall sometimes write dots in a memory sequence, [...A,B,...,C], to indicate the possible presence of previous memories which are irrelevant to the case being considered.)

Our problem is, then, to treat the interaction of such observer-systems with other physical systems (observations), within the framework of wave mechanics, and to deduce the resulting memory configurations, which we can then interpret as the subjective experiences of the observers.

We begin by defining what shall constitute a "good" observation. A good observation of a quantity A, with eigenfunctions $\{\phi_i\}$ for a system S, by an observer whose initial state is $\psi^O_{[\ldots]}$, shall consist of an interaction which, in a specified period of time, transforms each (total) state

$$\psi^{S+O} = \phi_i \psi^O_{[\ldots]}$$

into a new state

$$\psi^{S+O'} = \phi_i \psi^O_{i[\ldots,\alpha_i]} ,$$

where $\alpha_i$ characterizes the state $\phi_i$. (It might stand for a recording of the eigenvalue, for example.) That is, our requirement is that the system state, *if it is an eigenstate,* shall be unchanged, and that the observer state shall change so as to describe an observer that is "aware" of which eigenfunction it is, i.e., some property is recorded in the memory of the observer which characterizes $\phi_i$, such as the eigenvalue. The requirement that the eigenstates for the system be unchanged is necessary if the observation is to be significant (repeatable), and the requirement that the observer state change in a manner which is different for each eigenfunction is necessary if we are to be able to call the interaction an observation at all.

§2. *Deductions*

From these requirements we shall first deduce the result of an observation upon a system which is *not* in an eigenstate of the observation. We know, by our previous remark upon what constitutes a good observation that the interaction transforms states $\phi_i \psi^O_{[\ldots]}$ into states $\phi_i \psi^O_{i[\ldots,\alpha_i]}$. Consequently we can simply superpose these solutions of the wave equation to arrive at the final state for the case of an arbitrary initial system state. Thus if the initial system state is not an eigenstate, but a general state $\sum_i a_i \phi_i$, we get for the final total state:

$$\psi^{S+O} = \sum_i a_i \phi_i \psi^O_{i[\ldots,\alpha_i]} . \tag{2.1}$$

This remains true also in the presence of further systems which do not interact for the time of measurement. Thus, if systems $S_1, S_2, \ldots, S_n$ are present as well as O, with original states $\psi^{S_1}, \psi^{S_2}, \ldots, \psi^{S_n}$, and the only interaction during the time of measurement is between $S_1$ and O, the result of the measurement will be the transformation of the initial total state:

$$\psi^{S_1+S_2+\ldots+S_n+O} = \psi^{S_1} \psi^{S_2} \ldots \psi^{S_n} \psi^O_{[\ldots]}$$

into the final state:

$$\psi'^{S_1+S_2+\ldots+S_n+O} = \sum_i a_i \phi_i^{S_1} \psi^{S_2} \ldots \psi^{S_n} \psi^O_{i[\ldots,\alpha_i]} \tag{2.2}$$

where $a_i = \left(\phi_i^{S_1}, \psi^{S_1}\right)$ and $\phi_i^{S_1}$ are eigenfunctions of the observation.

Thus we arrive at the general rule for the transformation of total state functions which describe systems within which observation processes occur:

*Rule* 1. The observation of a quantity A, with eigenfunctions $\phi_i^{S_1}$, in a system $S_1$ by the observer O, transforms the total state according to:

$$\psi^{S_1}\psi^{S_2}\ldots\psi^{S_n}\psi^{O}_{[\ldots]} \rightarrow \sum_i a_i \phi_i^{S_1}\psi^{S_2}\ldots\psi^{S_n}\psi^{O}_{i[\ldots,\alpha_i]},$$

where $a_i = \left(\phi_i^{S_1}, \psi^{S_1}\right)$.

If we next consider a *second* observation to be made, where our total state is now a superposition, we can apply *Rule* 1 separately to each element of the superposition, since each element separately obeys the wave equation and behaves independently of the remaining elements, and then superpose the results to obtain the final solution. We formulate this as:

*Rule* 2. *Rule* 1 may be applied separately to each element of a superposition of total system states, the results being superposed to obtain the final total state. Thus, a determination of B, with eigenfunctions $\eta_j^{S_2}$, on $S_2$ by the observer O transforms the total state

$$\sum_i a_i \phi_i^{S_1}\psi^{S_2}\ldots\psi^{S_n}\psi^{O}_{i[\ldots,\alpha_i]}$$

into the state

$$\sum_{i,j} a_i b_j \phi_i^{S_1}\eta_j^{S_2}\psi^{S_3}\ldots\psi^{S_n}\psi^{O}_{ij[\ldots,\alpha_i,\beta_j]}$$

where $b_j = \left(\eta_j^{S_2}, \psi^{S_2}\right)$, which follows from the application of *Rule* 1 to each element $\phi_i^{S_1}\psi^{S_2}\ldots\psi^{S_n}\psi^{O}_{i[\ldots,\alpha_i]}$, and then superposing the results with the coefficients $a_i$.

These two rules, which follow directly from the superposition principle, give us a convenient method for determining final total states for any number of observation processes in any combinations. We must now seek the interpretation of such final total states.

Let us consider the simple case of a single observation of a quantity A, with eigenfunctions $\phi_i$, in the system S with initial state $\psi^S$, by an observer O whose initial state is $\psi^O_{[\ldots]}$. The final result is, as we have seen, the superposition:

$$\psi'^{S+O} = \sum_i a_i \phi_i \psi^O_{i[\ldots,\alpha_i]} \; . \tag{2.3}$$

We note that there is no longer any independent system state or observer state, although the two have become correlated in a one-one manner. However, in each *element* of the superposition (2.3), $\phi_i \psi^O_{i[\ldots,\alpha_i]}$, the object-system state is a particular eigenstate of the observer, and *furthermore the observer-system state describes the observer as definitely perceiving that particular system state.*[1] It is this correlation which allows one to maintain the interpretation that a measurement has been performed.

We now carry the discussion a step further and allow the observer-system to repeat the observation. Then according to *Rule* 2 we arrive at the total state after the second observation:

---

[1] At this point we encounter a language difficulty. Whereas before the observation we had a single observer state afterwards there were a number of different states for the observer, all occurring in a superposition. Each of these separate states is a state for an observer, so that we can speak of the different observers described by the different states. On the other hand, the same physical system is involved, and from this viewpoint it is the *same* observer, which is in different states for different elements of the superposition (i.e., has had different experiences in the separate elements of the superposition). In this situation we shall use the singular when we wish to emphasize that a single physical system is involved, and the plural when we wish to emphasize the different experiences for the separate elements of the superposition. (e.g., "The observer performs an observation of the quantity A, after which each of the observers of the resulting superposition has perceived an eigenvalue.")

(2.4) $$\psi'^{S+O} = \sum_i a_i \phi_i \psi^O_{ii[\ldots,\alpha_i,\alpha_i]} \ .$$

Again, we see that each element of (2.4), $\phi_i \psi^O_{ii[\ldots,\alpha_i,\alpha_i]}$, describes a system eigenstate, but this time also describes the observer as having obtained the *same result* for each of the two observations. Thus for every separate state of the observer in the final superposition, the result of the observation was repeatable, even though different for different states. This repeatability is, of course, a consequence of the fact that after an observation the *relative* system state for a particular observer state is the corresponding eigenstate.

Let us suppose now that an observer-system O, with initial state $\psi^O_{[\ldots]}$, measures the *same* quantity A in a number of separate identical systems which are initially in the same state, $\psi^{S_1} = \psi^{S_2} = \ldots = \psi^{S_n} = \sum_i a_i \phi_i$ (where the $\phi_i$ are, as usual, eigenfunctions of A). The initial total state function is then

(2.3) $$\psi_0^{S_1+S_2+\ldots+S_n+O} = \psi^{S_1} \psi^{S_2} \ldots \psi^{S_n} \psi^O_{[\ldots]} \ .$$

We shall assume that the measurements are performed on the systems in the order $S_1, S_2, \ldots, S_n$. Then the total state after the first measurement will be, by *Rule* 1,

(2.4) $$\psi_1^{S_1+S_2+\ldots+S_n+O} = \sum_i a_i \phi_i^{S_1} \psi^{S_2} \ldots \psi^{S_n} \psi^O_{i[\ldots,\alpha_i^1]}$$

(where $\alpha_i^1$ refers to the first system, $S_1$) .

After the second measurement it will be, by *Rule* 2,

(2.5) $$\psi_2^{S_1+S_2+\ldots S_n+O} = \sum_{i,j} a_i a_j \phi_i^{S_1} \phi_j^{S_2} \psi^{S_3} \ldots \psi^{S_n} \psi^O_{ij[\ldots,\alpha_i^1,\alpha_j^2]}$$

and in general, after $r$ measurements have taken place $(r \leq n)$ *Rule* 2 gives the result:

$$(2.6)\quad \psi_r = \sum_{i,j,\ldots,k} a_i a_j \ldots a_k \phi_i^{S_1} \phi_j^{S_2} \ldots \phi_k^{S_r} \psi^{S_{r+1}} \ldots \psi^{S_n} \psi^O_{ij\ldots k[\ldots,\alpha_i^1,\alpha_j^2,\ldots,\alpha_k^r]}$$

We can give this state, $\psi_r$, the following interpretation. It consists of a superposition of states:

$$(2.7)\quad \psi'_{ij\ldots k} = \phi_i^{S_1} \phi_j^{S_2} \ldots \phi_k^{S_r} \psi^{S_{r+1}} \ldots \psi^{S_n} \psi^O_{ij\ldots k[\ldots,\alpha_i^1,\alpha_j^2,\ldots,\alpha_k^r]}$$

each of which describes the observer with a definite memory sequence $[\ldots,\alpha_i^1,\alpha_j^2,\ldots,\alpha_k^r]$, and relative to whom the (observed) system states are the corresponding eigenfunctions $\phi_i^{S_1},\phi_j^{S_2},\ldots,\phi_k^{S_r}$, the remaining systems, $S_{r+1},\ldots S_n$, being unaltered.

In the language of subjective experience, the observer which is described by a typical element, $\psi'_{ij\ldots k}$, of the superposition has perceived an apparently random sequence of definite results for the observations. It is furthermore true, since in each element the system has been left in an eigenstate of the measurement, that if at this stage a redetermination of an earlier system observation $(S_\ell)$ takes place, every element of the resulting final superposition will describe the observer with a memory configuration of the form $[\ldots,\alpha_i^1,\ldots,\alpha_j^\ell,\ldots,\alpha_k^r,\alpha_j^\ell]$ in which the earlier memory coincides with the later – i.e., the memory states are *correlated*. It will thus *appear* to the observer which is described by a typical element of the superposition that each initial observation on a system caused the system to "jump" into an eigenstate in a random fashion and thereafter remain there for subsequent measurements on the same system. Therefore, qualitatively, at least, the probabilistic assertions of Process 1 *appear* to be valid to the observer described by a typical element of the final superposition.

In order to establish quantitative results, we must put some sort of measure (weighting) on the elements of a final superposition. This is necessary to be able to make assertions which will hold for almost all of the observers described by elements of a superposition. In order to make quantitative statements about the relative frequencies of the different possible results of observation which are recorded in the memory of a typical observer we must have a method of selecting a *typical* observer.

Let us therefore consider the search for a general scheme for assigning a measure to the elements of a superposition of orthogonal states $\sum a_i \phi_i$. We require then a positive function $\mathfrak{M}$ of the complex coefficients of the elements of the superposition, so that $\mathfrak{M}(a_i)$ shall be the measure assigned to the element $\phi_i$. In order that this general scheme shall be unambiguous we must first require that the states themselves always be normalized, so that we can distinguish the coefficients from the states. However, we can still only determine the *coefficients*, in distinction to the states, up to an arbitrary phase factor, and hence the function $\mathfrak{M}$ must be a function of the amplitudes of the coefficients alone, (i.e., $\mathfrak{M}(a_i) = \mathfrak{M}(\sqrt{a_i^* a_i})$ ), in order to avoid ambiguities.

If we now impose the additivity requirement that if we regard a subset of the superposition, say $\sum_{i=1}^{n} a_i \phi_i$, as a single element $\alpha \phi'$:

$$\alpha \phi' = \sum_{i=1}^{n} a_i \phi_i \ , \tag{2.8}$$

then the measure assigned to $\phi'$ shall be the sum of the measures assigned to the $\phi_i$ (i from 1 to n):

$$\mathfrak{M}(\alpha) = \sum_i \mathfrak{M}(a_i) \ , \tag{2.9}$$

then we have already restricted the choice of $\mathfrak{M}$ to the square amplitude alone. ($\mathfrak{M}(a_i) = a_i^* a_i$), apart from a multiplicative constant.)

To see this we note that the normality of $\phi'$ requires that $|\alpha| = \sqrt{\sum_{i=1}^{n} a_i^* a_i}$. From our remarks upon the dependence of $\mathfrak{M}$ upon the amplitude alone, we replace the $a_i$ by their amplitudes $\mu_i = |a_i|$.

(2.9) then requires that

$$(2.10)\quad \mathfrak{M}(a) = \mathfrak{M}\left(\sqrt{\sum a_i^* a_i}\right) = \mathfrak{M}\left(\sqrt{\sum \mu_i^2}\right) = \sum \mathfrak{M}(\mu_i) = \sum \mathfrak{M}(\sqrt{\mu_i^2}) \ .$$

Defining a new function $g(x)$:

$$(2.11)\qquad g(x) = \mathfrak{M}(\sqrt{x}) \ ,$$

we see that (2.10) requires that

$$(2.12)\qquad g\left(\sum \mu_i^2\right) = \sum g(\mu_i^2) \ ,$$

so that $g$ is restricted to be linear and necessarily has the form:

$$(2.13)\qquad g(x) = cx \qquad (c \text{ constant}) \ .$$

Therefore $g(x^2) = cx^2 = \mathfrak{M}\sqrt{x^2} = \mathfrak{M}(x)$ and we have deduced that $\mathfrak{M}$ is restricted to the form

$$(2.14)\qquad \mathfrak{M}(a_i) = \mathfrak{M}(\mu_i) = c\mu_i^2 = ca_i^* a_i \ ,$$

and we have shown that the only choice of measure consistent with our additivity requirement is the square amplitude measure, apart from an arbitrary multiplicative constant which may be fixed, if desired, by normalization requirements. (The requirement that the total measure be unity implies that this constant is 1.)

The situation here is fully analogous to that of classical statistical mechanics, where one puts a measure on trajectories of systems in the phase space by placing a measure on the phase space itself, and then making assertions which hold for "almost all" trajectories (such as ergodicity, quasi-ergodicity, etc).[2] This notion of "almost all" depends here also upon the choice of measure, which is in this case taken to be Lebesgue measure on the phase space. One could, of course, contradict

2 See Khinchin [16].

the statements of classical statistical mechanics by choosing a measure for which only the exceptional trajectories had nonzero measure. Nevertheless the choice of Lebesgue measure on the phase space can be justified by the fact that it is the only choice for which the "conservation of probability" holds, (Liouville's theorem) and hence the only choice which makes possible any reasonable statistical deductions at all.

In our case, we wish to make statements about "trajectories" of observers. However, for us a trajectory is constantly branching (transforming from state to superposition) with each successive measurement. To have a requirement analogous to the "conservation of probability" in the classical case, we demand that the measure assigned to a trajectory at one time shall equal the sum of the measures of its separate branches at a later time. This is precisely the additivity requirement which we imposed and which leads uniquely to the choice of square-amplitude measure. Our procedure is therefore quite as justified as that of classical statistical mechanics.

Having deduced that there is a unique measure which will satisfy our requirements, the square-amplitude measure, we continue our deduction. This measure then assigns to the $i,j,\dots,k^{th}$ element of the superposition (2.6),

$$\phi_i^{S_1}\phi_j^{S_2}\dots\phi_k^{S_r}\psi^{S_{r+1}}\dots\psi^{S_n}\psi^{O}_{ij\dots k[\dots,\alpha_i^1,\alpha_j^2,\dots,\alpha_k^r]} , \tag{2.15}$$

the measure (weight)

$$M_{ij\dots k} = (a_i a_j \dots a_k)^* (a_i a_j \dots a_k) , \tag{2.16}$$

so that the observer state with memory configuration $[\dots,\alpha_i^1,\alpha_j^2,\dots,\alpha_k^r]$ is assigned the measure $a_i^* a_i a_j^* a_j \dots a_k^* a_k = M_{ij\dots k}$. We see immediately that this is a product measure, namely

$$M_{ij\dots k} = M_i M_j \dots M_k , \tag{2.17}$$

where

$$M_\ell = a_\ell^* a_\ell ,$$

so that the measure assigned to a particular memory sequence $[\ldots,\alpha_i^1,\alpha_j^2,\ldots,\alpha_k^r]$ is simply the product of the measures for the individual components of the memory sequence.

We notice now a direct correspondence of our measure structure to the probability theory of random sequences. Namely, *if we were to regard* the $M_{ij\ldots k}$ as probabilities for the sequences $[\ldots,\alpha_i^1,\alpha_j^2,\ldots,\alpha_k^r]$, then the sequences are equivalent to the random sequences which are generated by ascribing to each term the *independent* probabilities $M_\ell = a_\ell^* a_\ell$. Now the probability theory is equivalent to measure theory mathematically, so that we can make use of it, while keeping in mind that all results should be translated back to measure theoretic language.

Thus, in particular, if we consider the sequences to become longer and longer (more and more observations performed) *each* memory sequence of the final superposition will satisfy any given criterion for a randomly generated sequence, generated by the independent probabilities $a_i^* a_i$, except for a set of total measure which tends toward zero as the number of observations becomes unlimited. Hence all averages of functions over *any* memory sequence, including the special case of frequencies, can be computed from the probabilities $a_i^* a_i$, except for a set of memory sequences of measure zero. We have therefore shown that the statistical assertions of Process 1 will appear to be valid to *almost all* observers described by separate elements of the superposition (2.6), in the limit as the number of observations goes to infinity.

While we have so far considered only sequences of observations of the same quantity upon identical systems, the result is equally true for arbitrary sequences of observations. For example, the sequence of observations of the quantities $A^1, A^2,\ldots,A^n,\ldots$ with (generally different) eigenfunction sets $\{\phi_i^1\}, \{\phi_j^2\},\ldots,\{\phi_k^n\},\ldots$ applied successively to the systems $S_1, S_2,\ldots,S_n,\ldots$, with (arbitrary) initial states $\psi^{S_1}, \psi^{S_2},\ldots,\psi^{S_n}$, ... transforms the total initial state:

$$\psi^{S_1+\ldots+S_n+O} = \psi^{S_1}\psi^{S_2}\ldots\psi^{S_n}\psi^O_{[\ldots]} \tag{2.18}$$

by rules 1 and 2, into the final state:

$$(2.19)\qquad \psi'^{S_1+S_2+\ldots+S_n+O} = \sum_{i,j,\ldots,k} (\phi_i^1,\psi^{S_1})(\phi_j^2,\psi^{S_2})\ldots(\phi_k^n,\psi^{S_n}) \ldots\phi_i^1\phi_j^2\ldots\phi_k^n\ldots\psi^O_{[\ldots,\alpha_i^1,\alpha_j^2,\ldots,\alpha_k^n,\ldots]},$$

where the memory sequence element $\alpha_\ell^r$ characterizes the $\ell^{th}$ eigenfunction, $\phi_\ell^r$ of the operator $A^r$. Again the square amplitude measure for each element of the superposition (2.19) reduces to the product measure of the individual memory element measures, $|(\phi_\ell^r,\psi^{S_r})|^2$ for the memory sequence element $\alpha_\ell^r$. Therefore, the memory sequence of a *typical* element of (2.19) has all the characteristics of a random sequence, with individual, independent (and now different), probabilities $|(\phi_\ell^r,\psi^{S_r})|^2$ for the $r^{th}$ memory state.

Finally, we can generalize to the case where several observations are allowed to be performed upon the *same* system. For example, if we permit the observation of a new quantity B, (eigenfunctions $\eta_m$, memory characterization $\beta_j$) upon the system $S_r$ for which $A^r$ has already been observed, then the state (2.19):

$$(2.20)\qquad \psi' = \sum_{i,\ell,\ldots,k} (\phi_i^1,\psi^{S_1})\ldots(\phi_\ell^r,\psi^{S_r})\ldots(\phi_k^n,\psi^{S_n}) \phi_i^1\ldots\phi_\ell^r\ldots\phi_k^n\ldots\psi^O_{[\ldots,\alpha_i^1,\ldots,\alpha_\ell^r,\ldots,\alpha_k^n,\ldots]}$$

is transformed by *Rule* 2 into the state:

$$(2.21)\qquad \psi' = \sum_{i,\ldots,\ell,\ldots,k,\underline{m}} (\phi_i^1,\psi^{S_1})\ldots(\phi_\ell^r,\psi^{S_r})\ldots(\phi_k^n,\psi^{S_n})\underline{(\eta_m^r,\phi_\ell^r)} \phi_i^1\ldots\phi_\mu^{r-1}\ldots\underline{\eta_m^r}\ldots\phi_\nu^{r+1}\ldots\phi_k^n\ldots\psi^O_{[\ldots,\alpha_i^1,\ldots,\alpha_\ell^r,\ldots,\alpha_k^n,\ldots,\underline{\beta_m^r}\ldots]}.$$

The *relative* system states for S have been changed from the eigenstates of $A^r, \{\phi_i^r\}$, to the eigenstates of $B^r, \{\eta_m^r\}$. We notice further that, with respect to our measure on the superposition, the memory sequences still have the character of random sequences, but of random sequences for which the individual terms are no longer independent. The memory states $\beta_m^r$ now depend upon the memory states $\alpha_\ell^r$ which represent the result of the previous measurement upon the same system, $S_r$. The *joint* (normalized) measure for this pair of memory states, conditioned by fixed values for remaining memory states is:

$$(2.22) \qquad M^{\alpha_i^1 \ldots \alpha_\mu^{r-1} \alpha_\nu^{r+1} \ldots \alpha_k^n}(\alpha_\ell^r, \beta_m^r) = \frac{M(\alpha_i^1, \ldots, \alpha_\ell^r, \ldots, \alpha_k^n, \beta_m^r)}{\sum_{\ell,m} M(\alpha_i^2, \ldots, \alpha_\ell^r, \ldots, \alpha_k^n, \beta_m^r)}$$

$$= \frac{|(\phi_i^1, \psi^{S_1}) \ldots (\phi_\ell^r, \psi^{S_r}) \ldots (\phi_k^n, \psi^{S_n})(\eta_m^r, \phi_\ell^r)|^2}{\sum_{\ell,m} |(\phi_i^1, \psi^{S_1}) \ldots (\phi_\ell^r, \psi^{S_r}) \ldots (\phi_k^n, \psi^{S_n})(\eta_m^r, \phi_\ell^r)|^2}$$

$$= |(\phi_\ell^r, \psi^{S_r})|^2 \, |(\eta_m^r, \phi_\ell^r)|^2 \ .$$

The joint measure (2.15) is, first of all, independent of the memory states for the remaining systems ($S_1 \ldots S_n$ excluding $S_r$). Second, the dependence of $\beta_m^r$ on $\alpha_\ell^r$ is *equivalent*, measure theoretically, to that given by the *stochastic process*[3] which converts the states $\phi_\ell^r$ into the states $\eta_m^r$ with transition probabilities:

$$(2.23) \qquad T_{\ell m} = \text{Prob.}\ (\phi_\ell^r \to \eta_m^r) = |(\eta_m^r, \phi_\ell^r)|^2 \ .$$

---

3 Cf. Chapter II, §6.

If we were to allow yet another quantity C to be measured in $S_r$, the new memory states $\alpha_p^r$ corresponding to the eigenfunctions of C would have a similar dependence upon the previous states $\beta_m^r$, but *no direct dependence* on the still earlier states $\alpha_\ell^r$. This dependence upon only the previous result of observation is a consequence of the fact that the *relative* system states are completely determined by the last observation.

We can therefore summarize the situation for an arbitrary sequence of observations, upon the same or different systems in any order, and for which the number of observations of each quantity in each system is very large, with the following result:

Except for a set of memory sequences of measure nearly zero, the averages of any functions over a memory sequence can be calculated approximately by the use of the independent probabilities given by Process 1 for each initial observation, on a system, and by the use of the transition probabilities (2.23) for succeeding observations upon the same system. In the limit, as the number of all types of observations goes to infinity the calculation is exact, and the exceptional set has measure zero.

This prescription for the calculation of averages over memory sequences by probabilities assigned to individual elements is precisely that of the orthodox theory (Process 1). Therefore all predictions of the usual theory will appear to be valid to the observer in almost all observer states, since these predictions hold for almost all memory sequences.

In particular, the uncertainty principle is never violated, since, as above, the latest measurement upon a system supplies all possible information about the relative system state, so that there is no direct correlation between any earlier results of observation on the system, and the succeeding observation. Any observation of a quantity B, between two successive observations of quantity A (all on the same system) will destroy the one-one correspondence between the earlier and later memory states for the result of A. Thus for alternating observations of different quantities there are fundamental limitations upon the correlations between memory states for the same observed quantity, these limitations expressing the content of the uncertainty principle.

In conclusion, we have described in this section processes involving an idealized observer, processes which are entirely deterministic and continuous from the over-all viewpoint (the total state function is presumed to satisfy a wave equation at all times) but whose result is a superposition, each element of which describes the observer with a different memory state. We have seen that in almost all of these observer states it *appears* to the observer that the probabilistic aspects of the usual form of quantum theory are valid. We have thus seen how pure wave mechanics, without any initial probability assertions, can lead to these notions on a subjective level, as appearances to observers.

### §3. *Several observers*

We shall now consider the consequences of our scheme when several observers are allowed to interact with the same systems, as well as with one another (communication). In the following discussion observers shall be denoted by $O_1, O_2, \ldots,$ other systems by $S_1, S_2, \ldots,$ and observables by operators A, B, C, with eigenfunctions $\{\phi_i\}$, $\{\eta_j\}$, $\{\xi_k\}$ respectively. The symbols $\alpha_i$, $\beta_j$, $\gamma_k$, occurring in memory sequences shall refer to characteristics of the states $\phi_i$, $\eta_j$, $\xi_k$, respectively. ($\psi^{O}_{i[\ldots,\alpha_i]}$ is interpreted as describing an observer, $O_j$, who has just observed the eigenvalue corresponding to $\phi_i$, i.e., who is "aware" that the system is in state $\phi_i$.)

We shall also wish to allow communication among the observers, which we view as an interaction by means of which the memory sequences of different observers become correlated. (For example, the transfer of impulses from the magnetic tape memory of one mechanical observer to that of another constitutes such a transfer of information.)[4] We shall regard these processes as observations made by one observer on another and shall use the notation that

---

4 We assume that such transfers merely duplicate, but do not destroy, the original information.

$$\psi^{O_j}_{i[\ldots,\alpha_i^{O_k}]}$$

represents a state function describing an observer $O_j$ who has obtained the information $\alpha_i$ from another observer, $O_k$. Thus the obtaining of information about A from $O_1$ by $O_2$ will transform the state

$$\psi^{O_1}_{i[\ldots,\alpha_i]}\psi^{O_2}_{[\ldots]}$$

into the state

$$\psi^{O_1}_{i[\ldots,\alpha_i]}\psi^{O_2}_{i[\ldots,\alpha_i^{1}]} \; . \tag{3.1}$$

*Rules* 1 and 2 are, of course, equally applicable to these interactions. We shall now illustrate the possibilities for several observers, by considering several cases.

*Case* 1: We allow two observers to separately observe the same quantity in a system, and then compare results.

We suppose that first observer $O_1$ observes the quantity A for the system S. Then by *Rule* 1 the original state

$$\psi^{S+O_1+O_2} = \psi^S\psi^{O_1}_{[\ldots]}\psi^{O_2}_{[\ldots]}$$

is transformed into the state

$$\psi' = \sum_i (\phi_i^S,\psi^S)\phi_i^S\psi^{O_1}_{i[\ldots,\alpha_i]}\psi^{O_2}_{[\ldots]} \; . \tag{3.2}$$

We now suppose that $O_2$ observes A, and by *Rule* 2 the state becomes:

$$\psi'' = \sum_i (\phi_i^S,\psi^S)\phi_i^S\psi^{O_1}_{i[\ldots,\alpha_i]}\psi^{O_2}_{i[\ldots,\alpha_i]} \; . \tag{3.3}$$

We now allow $O_2$ to "consult" $O_1$, which leads in the same fashion from (3.1) and *Rule* 2 to the final state

$$\psi''' = \sum_i (\phi_i^S, \psi^S)\phi_i^S \psi^{O_1}_{i[\ldots,a_i]} \psi^{O_2}_{ii[\ldots,a_i,a_i^{O_1}]} \quad (3.4)$$

Thus, for every element of the superposition the information obtained from $O_1$ agrees with that obtained directly from the system. This means that observers who have separately observed the same quantity will *always* agree with each other.

Furthermore, it is obvious at this point that the same result, (4.4), is obtained if $O_2$ *first* consults $O_1$, then performs the direct observation, except that the memory sequence for $O_2$ is reversed ($[\ldots,a_i^{O_1},a_i]$ instead of $[\ldots,a_i,a_i^{O_1}]$). There is still perfect agreement in every element of the superposition. Therefore, information obtained from another observer is always reliable, since subsequent direct observation will always verify it. We thus see the central role played by correlations in wave functions for the preservation of consistency in situations where several observers are allowed to consult one another. It is the transitivity of correlation in these cases (that if $S_1$ is correlated to $S_2$, and $S_2$ to $S_3$, then so is $S_1$ to $S_2$) which is responsible for this consistency.

*Case* 2: We allow two observers to measure separately two different, non-commuting quantities in the same system.

Assume that first $O_1$ observes A for the system, so that, as before, the initial state $\psi^S \psi^{O_1} \psi^{O_2}$ is transformed to:

$$\psi' = \sum_i (\phi_i, \psi^S)\phi_i \psi^{O_1}_{i[\ldots,a_i]} \psi^{O_2}_{[\ldots]} \quad (3.5)$$

Next let $O_2$ determine $\beta$ for the system, where $\{\eta_j\}$ are the eigenfunctions of $\beta$. Then by application of *Rule* 2 the result is

(3.6) $$\psi'' = \sum_{i,j} (\phi_i, \psi^S)(\eta_j, \phi_i)(\eta_j \psi^{O_1}_{i[\ldots,a_i]} \psi^{O_2}_{j[\ldots,\beta_j]}$$

$O_2$ is now perfectly correlated with the system, since a redetermination by him will lead to agreeing results. This is no longer the case for $O_1$, however, since a redetermination of A by him will result in (by *Rule* 2)

(3.7) $$\psi'' = \sum_{i,j,k} (\phi_i, \psi^S)(\eta_j, \phi_i)(\phi_k, \eta_j)\phi^S_k \psi^{O_2}_{j[\ldots,\beta_j]} \psi^{O_1}_{ik[\ldots,a_i,a_k]} \cdot$$

Hence the second measurement of $O_1$ does not in all cases agree with the first, and has been upset by the intervention of $O_2$.

We can deduce the statistical relation between $O_1$'s first and second results ($a_i$ and $a_k$) by our previous method of assigning a measure to the elements of the superposition (3.7). The measure assigned to the $(i, j, k)^{th}$ element is then:

(3.8) $$M_{ijk} = |(\phi_i, \psi^S)(\eta_j, \phi_i)(\phi_k, \eta_j)|^2 \ .$$

This measure is equivalent, in this case, to the probabilities assigned by the orthodox theory (Process 1), where $O_2$'s observation is regarded as having converted each state $\phi_i$ into a non-interfering mixture of states $\eta_j$, weighted with probabilities $|(\eta_j, \phi_i)|^2$, upon which $O_1$ makes his second observation.

Note, however, that this equivalence with the statistical results obtained by considering that $O_2$'s observation changed the system state into a mixture, holds true *only so long as* $O_1$'s *second observation is restricted to the system.* If he were to attempt to simultaneously determine a property of the system as well as of $O_2$, interference effects might become important. The description of the states relative to $O_1$, after $O_2$'s observation, as non-interfering mixtures is therefore incomplete.

*Case* 3: We suppose that two systems $S_1$ and $S_2$ are correlated but no longer interacting, and that $O_1$ measures property A in $S_1$, and $O_2$ property $\beta$ in $S_2$.

We wish to see whether $O_2$'s intervention with $S_2$ can in any way affect $O_1$'s results in $S_1$, so that perhaps signals might be sent by these means. We shall assume that the initial state for the system pair is

$$\psi^{S_1+S_2} = \sum_i a_i \phi_i^{S_1} \phi_i^{S_2} . \tag{3.9}$$

We now allow $O_1$ to observe A in $S_1$, so that after this observation the total state becomes:

$$\psi'^{S_1+S_2+O_1+O_2} = \sum_i a_i \phi_i^{S_1} \phi_i^{S_2} \psi_{i[\ldots,\alpha_i]}^{O_1} \psi_{[\ldots]}^{O_2} . \tag{3.10}$$

$O_1$ can of course continue to repeat the determination, obtaining the same result each time.

We now suppose that $O_2$ determines $\beta$ in $S_2$, which results in

$$\psi'' = \sum_{i,j} a_i (\eta_j^2, \phi_i^2) \phi_i^1 \eta_j^2 \psi_{i[\ldots,\alpha_i]}^{O_1} \psi_{j[\ldots,\beta_j]}^{O_2} . \tag{3.11}$$

However, in this case, as distinct from *Case* 2, we see that the intervention of $O_2$ in no way affects $O_1$'s determinations, since $O_1$ is still perfectly correlated to the states $\phi_i^{S_1}$ of $S_1$, and any further observations by $O_1$ will lead to the same results as the earlier observations. Thus each memory sequence for $O_1$ continues without change due to $O_2$'s observation, and such a scheme could not be used to send any signals.

Furthermore, we see that the result (3.11) is arrived at even in the case that $O_2$ should make his determination before that of $O_1$. Therefore any expectations for the outcome of $O_1$'s first observation are in no way affected by whether or not $O_2$ performs his observation before that of $O_1$. This is true because the expectation of the outcome for $O_1$ can be computed from (4.10), which is the same whether or not $O_2$ performs his measurement before or after $O_1$.

It is therefore seen that one observer's observation upon one system of a correlated, but non-interacting pair of systems, has no effect on the remote system, in the sense that the outcome or expected outcome of any experiments by another observer on the remote system are not affected. Paradoxes like that of Einstein-Rosen-Podolsky[5] which are concerned with such correlated, non-interacting, systems are thus easily understood in the present scheme.

Many further combinations of several observers and systems can be easily studied in the present framework, and all questions answered by first writing down the final state for the situation with the aid of the *Rules* 1 and 2, and then noticing the relations between the elements of the memory sequences.

---

5 Einstein [8].

# V. SUPPLEMENTARY TOPICS

We have now completed the abstract treatment of measurement and observation, with the deduction that the statistical predictions of the usual form of quantum theory (Process 1) will appear to be valid to all observers. We have therefore succeeded in placing our theory in correspondence with experience, at least insofar as the ordinary theory correctly represents experience.

We should like to emphasize that this deduction was carried out by using only the principle of superposition, and the postulate that an observation has the property that *if* the observed variable has a definite value in the object-system then it will remain definite and the observer will perceive this value. This treatment is therefore valid for any possible quantum interpretation of observation processes, i.e., any way in which one can interpret wave functions as describing observers, as well as for any form of quantum mechanics for which the superposition principle for states is maintained. Our abstract discussion of observation is therefore logically complete, in the sense that our results for the subjective experience of observers are correct, if there are any observers at all describable by wave mechanics.[1]

In this chapter we shall consider a number of diverse topics from the point of view of our pure wave mechanics, in order to supplement the abstract discussion and give a feeling for the new viewpoint. Since we are now mainly interested in elucidating the reasonableness of the theory, we shall often restrict ourselves to plausibility arguments, rather than detailed proofs.

---

1 They are, of course, vacuously correct otherwise.

§1. *Macroscopic objects and classical mechanics*

In the light of our knowledge about the atomic constitution of matter, any "object" of macroscopic size is composed of an enormous number of constituent particles. The wave function for such an object is then in a space of fantastically high dimension (3N, if N is the number of particles). Our present problem is to understand the existence of macroscopic objects, and to relate their ordinary (classical) behavior in the three dimensional world to the underlying wave mechanics in the higher dimensional space.

Let us begin by considering a relatively simple case. Suppose that we place in a box an electron and a proton, each in a definite momentum state, so that the position amplitude density of each is uniform over the whole box. After a time we would expect a hydrogen atom in the ground state to form, with ensuing radiation. We notice, however, that the position amplitude density of each particle is *still* uniform over the whole box. Nevertheless the amplitude distributions are now no longer independent, but correlated. In particular, the *conditional* amplitude density for the electron, conditioned by any definite proton (or centroid) position, is *not* uniform, but is given by the familiar ground state wave function for the hydrogen atom. What we mean by the statement, "a hydrogen atom has formed in the box," is just that this correlation has taken place – a correlation which insures that the *relative* configuration for the electron, for a definite proton position, conforms to the customary ground state configuration.

The wave function for the hydrogen atom can be represented as a product of a centroid wave function and a wave function over relative coordinates, where the centroid wave function obeys the wave equation for a particle with mass equal to the total mass of the proton-electron system. Therefore, if we now open our box, the centroid wave function will spread with time in the usual manner of wave packets, to eventually occupy a vast region of space. The *relative* configuration (described by the *relative coordinate* state function) has, however, a permanent nature, since it represents a bound state, and it is this relative configuration which we usually think of as the object called the hydrogen atom. Therefore, no matter how indefinite the positions of the individual particles become in the total state function (due to the spreading of the centroid), this state can be regarded as giving (through the centroid wave function) an amplitude distribution over a comparatively definite object, the tightly bound electron-proton system. The general state, then, does not describe any single such definite object, but a superposition of such cases with the object located at different positions.

In a similar fashion larger and more complex objects can be built up through strong correlations which bind together the constituent particles. It is still true that the general state function for such a system may lead to marginal position densities for any single particle (or centroid) which extend over large regions of space. Nevertheless we can speak of the existence of a relatively definite object, since the specification of a single position for a particle, or the centroid, leads to the case where the *relative* position densities of the remaining particles are distributed closely about the specified one, in a manner forming the comparatively definite object spoken of.

Suppose, for example, we begin with a cannonball located at the origin, described by a state function:

$$\psi_{[c_j(0,0,0)]} ,$$

where the subscript indicates that the total state function $\psi$ describes a system of particles bound together so as to form an object of the size and shape of a cannonball, whose centroid is located (approximately) at the origin, say in the form of a real gaussian wave packet of small dimensions, with variance $\sigma_0^2$ for each dimension.

If we now allow a long lapse of time, the centroid of the system will spread in the usual manner to occupy a large region of space. (The spread in each dimension after time t will be given by $\sigma_t^2 = \sigma_0^2 + (\hbar^2 t^2/4\sigma_0^2 m^2)$, where m is the mass.) Nevertheless, for any *specified* centroid position, the particles, since they remain in bound states, have distributions which again correspond to the fairly well defined size and shape of the cannonball. Thus the total state can be regarded as a (continuous) superposition of states

$$\psi = \int a_{xyz}\, \psi_{[c_j(x,y,z)]}\, dx\, dy\, dz \ ,$$

*each of which* ($\psi_{[c_j(x,y,z)]}$) describes a cannonball at the position $(x, y, z)$. The coefficients $a_{xyz}$ of the superposition then correspond to the centroid distribution.

It is *not* true that each individual particle spreads independently of the rest, in which case we would have a final state which is a grand superposition of states in which the particles are located independently everywhere. The fact that they are in bound states restricts our final state to a superposition of "cannonball" states. The wave function for the centroid can therefore be taken as a representative wave function for the whole object.

It is thus in this sense of correlations between constituent particles that definite macroscopic objects can exist within the framework of pure wave mechanics. The building up of correlations in a complex system supplies us with a mechanism which also allows us to understand how condensation phenomena (the formation of spatial boundaries which separate phases of different physical or chemical properties) can be controlled by the wave equation, answering a point raised by Schrödinger

Classical mechanics, also, enters our scheme in the form of correlation laws. Let us consider a system of objects (in the previous sense), such that the centroid of each object has initially a fairly well defined position and momentum (e.g., let the wave function for the centroids consist of a product of gaussian wave packets). As time progresses, the centers of the square amplitude distributions for the objects will move in a manner approximately obeying the laws of motion of classical mechanics, with the degree of approximation depending upon the masses and the length of time considered, as is well known. (Note that we do not mean to imply that the wave packets of the individual objects remain independent if they are interacting. They do not. The motion that we refer to is that of the centers of the *marginal* distributions for the centroids of the bodies.)

The general state of a system of macroscopic objects does not, however, ascribe any nearly definite positions and momenta to the individual bodies. Nevertheless, any general state can at any instant be analyzed into a *superposition* of states each of which *does* represent the bodies with fairly well defined positions and momenta.[2] Each of these states then propagates approximately according to classical laws, so that the general state can be viewed as a superposition of quasi-classical states propagating according to nearly classical trajectories. In other words, if the masses are large or the time short, there will be strong correlations between the initial (approximate) positions and momenta and those at a later time, with the dependence being given approximately by classical mechanics.

Since large scale objects obeying classical laws have a place in our theory of pure wave mechanics, we have justified the introduction of

---

[2] For any $\varepsilon$ one can construct a complete orthonormal set of (one particle) states $\phi_{\mu,\nu}$, where the double index $\mu,\nu$ refers to the approximate position and momentum, and for which the expected position and momentum values run independently through sets of approximately uniform density, such that the position and momentum uncertainties, $\sigma_x$ and $\sigma_p$, satisfy $\sigma_x \leqq C\varepsilon$ and $\sigma_p \leqq C\frac{\hbar}{2\varepsilon}$ for each $\phi_{\mu,\nu}$, where C is a constant $\sim 60$. The uncertainty product then satisfies $\sigma_x\sigma_p \leqq C^2\frac{\hbar}{2}$, about 3,600 times the minimum allowable, but still sufficiently low for macroscopic objects. This set can then be used as a basis for our decomposition into states where every body has a roughly defined position and momentum. For a more complete discussion of this set see von Neumann [17], pp. 406-407.

models for observers consisting of classically describable, automatically functioning machinery, and the treatment of observation of Chapter IV is non-vacuous.

Let us now consider the result of an observation (considered along the lines of Chapter IV) performed upon a system of macroscopic bodies in a general state. The observer will *not* become aware of the fact that the state does not correspond to definite positions and momenta (i.e., he will not see the objects as "smeared out" over large regions of space) but will himself simply become correlated with the system – after the observation the composite system of objects + observer will be in a superposition of states, each element of which describes an observer who has perceived that the objects have nearly definite positions and momenta, and for whom the relative system state is a quasi-classical state in the previous sense, and furthermore to whom the system will appear to behave according to classical mechanics if his observation is continued. We see, therefore, how the classical appearance of the macroscopic world to us can be explained in the wave theory.

§2. *Amplification processes*

In Chapter III and IV we discussed abstract measuring processes, which were considered to be simply a direct coupling between two systems, the object-system and the apparatus (or observer). There is, however, in actuality a whole chain of intervening systems linking a microscopic system to a macroscopic observer. Each link in the chain of intervening systems becomes correlated to its predecessor, so that the result is an amplification of effects from the microscopic object-system to a macroscopic apparatus, and then to the observer.

The amplification process depends upon the ability of the state of one micro-system (particle, for example) to become correlated with the states of an enormous number of other microscopic systems, the totality of which we shall call a detection system. For example, the totality of gas atoms in a Geiger counter, or the water molecules in a cloud chamber, constitute such a detection system.

The amplification is accomplished by arranging the condition of the detection system so that the states of the individual micro-systems of the detector are *metastable*, in a way that if one micro-system should fall from its metastable state it would influence the reduction of others. This type of arrangement leaves the entire detection system metastable against chain reactions which involve a large number of its constituent systems. In a Geiger counter, for example, the presence of a strong electric field leaves the gas atoms metastable against ionization. Furthermore, the products of the ionization of one gas atom in a Geiger counter can cause further ionizations, in a cascading process. The operation of cloud chambers and photographic films is also due to metastability against such chain reactions.

The chain reactions cause large numbers of the micro-systems of the detector to behave as a unit, all remaining in the metastable state, or all discharging. In this manner the states of a sufficiently large number of micro-systems are correlated, so that one can speak of the whole ensemble being in a state of discharge, or not.

For example, there are essentially only two macroscopically distinguishable states for a Geiger counter; discharged or undischarged. The correlation of large numbers of gas atoms, due to the chain reaction effect, implies that either very few, or else very many of the gas atoms are ionized at a given time. Consider the complete state function $\psi^G$ of a Geiger counter, which is a function of all the coordinates of all of the constituent particles. Because of the correlation of the behavior of a large number of the constituent gas atoms, the total state $\psi^G$ can always be written as a superposition of two states

$$\psi^G = a_1 \psi^1_{[U]} + a_2 \psi^2_{[D]} , \tag{2.1}$$

where $\psi^1_{[U]}$ signifies a state where only a small number of gas atoms are ionized, and $\psi^2_{[D]}$ a state for which a large number are ionized.

To see that the decomposition (2.1) is valid, expand $\psi^G$ in terms of individual gas atom stationary states:

$$\psi^G = \sum_{i,j,\ldots,k} a_{ij\ldots k} \psi_i^{S_1} \psi_j^{S_2} \ldots \psi_k^{S_n} , \tag{2.2}$$

where $\psi_\ell^{S_r}$ is the $\ell^{th}$ state of atom r. Each element of the superposition (2.2)

$$\psi_i^{S_1} \psi_j^{S_2} \ldots \psi_k^{S_n} \tag{2.3}$$

must contain either a very large number of atoms in ionized states, or else a very small number, because of the chain reaction effect. By choosing some medium-sized number as a dividing line, each element of (2.2) can be placed in one of the two categories, high number of low number of ionized atoms. If we then carry out the sum (2.2) over only those elements of the first category, we get a state (and coefficient)

$$a_1 \psi_{[D]}^1 = \sum_{ij\ldots k}{}' a_{ij\ldots k} \psi_i^{S_1} \psi_j^{S_2} \ldots \psi_k^{S_n} . \tag{2.4}$$

The state $\psi_{[D]}^1$ is then a state where a large number of particles are ionized. The subscript [D] indicates that it describes a Geiger counter which has discharged. If we carry out the sum over the remaining terms of (2.2) we get in a similar fashion:

$$a_2 \psi_{[U]}^2 = \sum_{ij\ldots k}{}'' a_{ij\ldots k} \psi_i^{S_1} \psi_j^{S_2} \ldots \psi_k^{S_n} \tag{2.5}$$

where [U] indicates the undischarged condition. Combining (2.4) and (2.5) we arrive at the desired relation (2.1). So far, this method of decomposition can be applied to any system, whether or not it has the chain reaction property. However, in our case, more is implied, namely that the spread of the number of ionized atoms in both $\psi_{[D]}$ and $\psi_{[U]}$ will be small compared to the separation of their averages, due to the fact that the existence of the chain reactions means that either many or else few atoms will be ionized, with the middle ground virtually excluded.

This type of decomposition is also applicable to all other detection devices which are based upon this chain reaction principle (such as cloud chambers, photo plates, etc.).

We consider now the coupling of such a detection device to another micro-system (object-system) for the purpose of measurement. If it is true that the initial object-system state $\phi_1$ will at some time t trigger the chain reaction, so that the state of the counter becomes $\psi^1_{[D]}$, while the object-system state $\phi_2$ will not, then it is still true that the initial object-system state $a_1\phi_1 + a_2\phi_2$ will result in the superposition

$$a_1\phi_1' \psi^1_{[D]} + a_2\phi_2' \psi^2_{[U]} \tag{2.6}$$

at time t.

For example, let us suppose that a particle whose state is a wave packet $\phi$, of linear extension greater than that of our Geiger counter, approaches the counter. Just before it reaches the counter, it can be decomposed into a superposition $\phi = a_1\phi_1 + a_2\phi_2$ ($\phi_1, \phi_2$ orthogonal) where $\phi_1$ has non-zero amplitude only in the region before the counter and $\phi_2$ has non-zero amplitude elsewhere (so that $\phi_1$ is a packet which will entirely pass through the counter while $\phi_2$ will entirely miss the counter). The initial total state for the system particle + counter is then:

$$\phi\psi_{[U]} = (a_1\phi_1 + a_2\phi_2)\psi_{[U]} \ ,$$

where $\psi_{[U]}$ is the initial (assumed to be undischarged) state of the counter.

But at a slightly later time $\phi_1$ is changed to $\phi_1'$, after traversing the counter and causing it to go into a discharged state $\psi^1_{[D]}$, while $\phi_2$ passes by into a state $\phi_2'$ leaving the counter in an undischarged state $\psi^2_{[U]}$. Superposing these results, the total state at the later time is

$$a_1\phi_1'\psi^1_{[D]} + a_2\phi_2'\psi^2_{[U]} \tag{2.7}$$

in accordance with (2.6). Furthermore, the relative particle state for $\psi^1_{[D]}$, $\phi_1'$, is a wave packet emanating from the counter, while the relative state for $\psi^2_{[U]}$ is a wave with a "shadow" cast by the counter. The counter therefore serves as an apparatus which performs an approximate position measurement on the particle.

No matter what the complexity or exact mechanism of a measuring process, the general superposition principle as stated in Chapter III, §3, remains valid, and our abstract discussion is unaffected. It is a vain hope that somewhere embedded in the intricacy of the amplification process is a mechanism which will somehow prevent the macroscopic apparatus state from reflecting the same indefiniteness as its object-system.

## §3. *Reversibility and irreversibility*

Let us return, for the moment, to the probabilistic interpretation of quantum mechanics based on Process 1 as well as Process 2. Suppose that we have a large number of identical systems (ensemble), and that the $j^{th}$ system is in the state $\psi^j$. Then for purposes of calculating expectation values for operators over the ensemble, the ensemble is represented by the mixture of states $\psi^j$ weighted with $1/N$, where $N$ is the number of systems, for which the density operator[3] is:

$$\rho = \frac{1}{N}\sum_j [\psi^j] , \tag{3.1}$$

where $[\psi^j]$ denotes the projection operator on $\psi^j$. This density operator, in turn, is equivalent to a density operator which is a sum of projections on orthogonal states (the eigenstates of $\rho$):[4]

---

3 Cf. Chapter III, §1.

4 See Chapter III, §2, particularly footnote 6, p. 46.

$$(3.2)\qquad \rho = \sum_i P_i\,[\eta_i]\ , \quad (\eta_i, \eta_j) = \delta_{ij},\ \sum_i P_i = 1\ ,$$

so that any ensemble is always equivalent to a mixture of orthogonal states, which representation we shall henceforth assume.

Suppose that a quantity A, with (non-degenerate) eigenstates $\{\phi_j\}$ is measured in each system of the ensemble. This measurement has the effect of transforming each state $\eta_i$ into the state $\phi_j$, with probability $|(\phi_j, \eta_i)|^2$; i.e., it will transform a large ensemble of systems in the state $\eta_i$ into an ensemble represented by the mixture whose density operator is $\sum_j |(\phi_j, \eta_i)|^2\,[\phi_j]$. Extending this result to the case where the original ensemble is a mixture of the $\eta_i$ weighted by $P_i$ ((3.2)), we find that the density operator $\rho$ is transformed by the measurement of A into the new density operator $\rho'$:

$$(3.3)\quad \rho' = \sum_i P_i \sum_j |(\eta_i, \phi_j)|^2\,[\phi_j] = \sum_j \Big(\sum_i P_i(\phi_j, (\eta_i, \phi_j)\eta_i)\Big)[\phi_j]$$

$$= \sum_j \Big(\phi_j, \sum_i P_i[\eta_i]\phi_j\Big)[\phi_j] = \sum_j (\phi_j, \rho\phi_j)[\phi_j]\ .$$

This is the general law by which mixtures change through Process 1.

However, even when no measurements are taking place, the states of an ensemble are changing according to Process 2, so that after a time interval t each state $\psi$ will be transformed into a state $\psi' = U_t\psi$, where $U_t$ is a unitary operator. This natural motion has the consequence that each mixture $\rho = \sum_i P_i[\eta_i]$ is carried into the mixture $\rho' = \sum_i P_i[U_t\eta_i]$ after a time t. But for every state $\xi$,

$$(3.4)\qquad \rho'\xi = \sum_i P_i[U_t\eta_i]\xi = \sum_i P_i(U_t\eta_i, \xi)U_t\eta_i$$

$$= U_t\sum_i P_i(\eta_i, U_t^{-1}\xi)\eta_i = U_t\sum_i P_i[\eta_i](U_t^{-1}\xi)$$

$$= (U_t\rho U_t^{-1})\xi\ .$$

Therefore

(3.5) $$\rho' = U_t \rho U_t^{-1} ,$$

which is the general law for the change of a mixture according to Process 2.

We are now interested in whether or not we get from any mixture to another by means of these two processes, i.e., if for any pair $\rho, \rho'$, there exist quantities A which can be measured and unitary (time dependence) operators U such that $\rho$ can be transformed into $\rho'$ by suitable applications of Processes 1 and 2. We shall see that this is not always possible, and that Process 1 can cause irreversible changes in mixtures.

For each mixture $\rho$ we define a quantity $I_\rho$:

(3.6) $$I_\rho = \text{Trace}\,(\rho \ln \rho) .$$

This number, $I_\rho$, has the character of information. If $\rho = \sum_i P_i [\eta_i]$, a mixture of orthogonal states $\eta_i$ weighted with $P_i$, then $I_\rho$ is simply the information of the distribution $P_i$ over the eigenstates of $\rho$ (relative to the uniform measure). (Trace $(\rho \ln \rho)$ is a unitary invariant and is proportional to the negative of the entropy of the mixture, as discussed in Chapter III, §2.)

Process 2 therefore has the property that it leaves $I_\rho$ unchanged, because

(3.7) $$I_{\rho'} = \text{Trace}\,(\rho' \ln \rho') = \text{Trace}\,(U_t \rho U_t^{-1} \ln U_t \rho U_t^{-1})$$

$$= \text{Trace}\,(U_t \rho \ln \rho U_t^{-1}) = \text{Trace}\,(\rho \ln \rho) = I_\rho .$$

Process 1, on the other hand, can decrease $I_\rho$ but never increase it. According to (3.3):

(3.8) $$\rho' = \sum_j (\phi_j, \rho\phi_j)[\phi_j] = \sum_{i,j} P_i\, |(\eta_i, \phi_j)|^2 [\phi_j] = \sum_j P'_j [\phi_j] ,$$

where $\rho'_j \quad \sum_i P_i T_{ij}$ and $T_{ij} = |(\eta_i, \phi_j)|^2$ is a doubly-stochastic matrix.[5] But $I_{\rho'} = \sum_j P'_j \ln P'_j$ and $I_\rho = \sum_i P_i \ln P_i$, with the $P_i, P'_j$ connected by $T_{ij}$, implies, by the theorem of information decrease for stochastic processes (II-§6), that:

$$I_{\rho'} \leq I_\rho \,. \tag{3.9}$$

Moreover, it can easily be shown by a slight strengthening of the theorems of Chapter II, §6 that *strict* inequality must hold unless (for each $i$ such that $\rho_i > 0$) $T_{ij} = 1$ for one $j$ and $0$ for the rest ($T_{ij} = \delta_{ik_j}$). This means that $|(\eta_i, \phi_j)|^2 = \delta_{ik_j}$, which implies that the original mixture was already a mixture of eigenstates of the measurement.

We have answered our question, and it is *not* possible to get from any mixture to another by means of Processes 1 and 2. There is an essential irreversibility to Process 1, since it corresponds to a stochastic process, which cannot be compensated by Process 2, which is reversible, like classical mechanics.[6]

Our theory of pure wave mechanics, to which we now return, must give equivalent results on the subjective level, since it leads to Process 1 there. Therefore, measuring processes will appear to be irreversible to any observers (even though the composite system including the observer changes its state reversibly).

---

5 Since $\sum_i T_{ij} = \sum_i |(\eta_i, \phi_j)|^2 = \sum_i (\phi_j, [\eta_i]\phi_j) = (\phi_j, \sum_i [\eta_i]\phi_j) = (\phi_j, I\phi_j) = 1$, and similarly $\sum_j T_{ij} = 1$ because $T_{ij}$ is symmetric.

6 For another, more complete, discussion of this topic in the probabilistic interpretation see von Neumann [17], Chapter V, §4.

There is another way of looking at this apparent irreversibility within our theory which recognizes only Process 2. When an observer performs an observation the result is a superposition, each element of which describes an observer who has perceived a particular value. From this time forward there is no interaction between the separate elements of the superposition (which describe the observer as having perceived different results), since each element separately continues to obey the wave equation. Each observer described by a particular element of the superposition behaves in the future completely independently of any events in the remaining elements, and he can no longer obtain any information whatsoever concerning these other elements (they are completely unobservable to him).

The irreversibility of the measuring process is therefore, within our framework, simply a subjective manifestation reflecting the fact that in observation processes the state of the observer is transformed into a superposition of observer states, each element of which describes an observer who is irrevocably cut off from the remaining elements. While it is conceivable that some outside agency could reverse the total wave function, such a change cannot be brought about by any observer which is represented by a single element of a superposition, since he is entirely powerless to have any influence on any other elements.

There are, therefore, fundamental restrictions to the knowledge that an observer can obtain about the state of the universe. It is impossible for any observer to discover the total state function of any physical system, since the process of observation itself leaves no independent state for the system or the observer, but only a composite system state in which the object-system states are inextricably bound up with the observer states. As soon as the observation is performed, the composite state is split into a superposition for which each element describes a different object-system state and an observer with (different) knowledge of it. Only the totality of these observer states, with their diverse knowledge, contains complete information about the original object-system state — but there is no possible communication between the observers described by these separate states. Any single observer can therefore possess knowledge only of the relative state function (relative to his state) of any systems, which is in any case all that is of any importance to him.

We conclude this section by commenting on another question which might be raised concerning irreversible processes: Is it necessary for the existence of measuring apparata, which can be correlated to other systems, to have frictional processes which involve systems of a large number of degrees of freedom? Are such thermodynamically irreversible processes possible in the framework of pure wave mechanics with a reversible wave equation, and if so, does this circumstance pose any difficulties for our treatment of measuring processes?

In the first place, it is certainly not necessary for dissipative processes involving additional degrees of freedom to be present before an interaction which correlates an apparatus to an object-system can take place. The counter-example is supplied by the simplified measuring process of III-§3, which involves only a system of one coordinate and an apparatus of one coordinate and no further degrees of freedom.

To the question whether such processes are possible within reversible wave mechanics, we answer *yes*, in the same sense that they are present in classical mechanics, where the microscopic equations of motion are also reversible. This type of irreversibility, which might be called *macroscopic irreversibility*, arises from a failure to separate "macroscopically indistinguishable" states into "true" microscopic states.[7] It has a fundamentally different character from the irreversibility of Process 1, which applies to micro-states as well and is peculiar to quantum mechanics. Macroscopically irreversible phenomena are common to both classical and quantum mechanics, since they arise from our incomplete information concerning a system, not from any intrinsic behavior of the system.[8]

---

7 See any textbook on statistical mechanics, such as ter Haar [11], Appendix I.

8 Cf. the discussion of Chapter II, §7. See also von Neumann [17], Chapter V, §4.

Finally, even when such frictional processes are involved, they present no new difficulties for the treatment of measuring and observation processes given here. We imposed no restrictions on the complexity or number of degrees of freedom of measuring apparatus or observers, and if any of these processes are present (such as heat reservoirs, etc.) then these systems are to be simply included as part of the apparatus or observer.

## §4. *Approximate measurement*

A phenomenon which is difficult to understand within the framework of the probabilistic interpretation of quantum mechanics is the result of an approximate measurement. In the abstract formulation of the usual theory there are two fundamental processes; the discontinuous, probabilistic Process 1 corresponding to precise measurement, and the continuous, deterministic Process 2 corresponding to absence of any measurement. What mixture of probability and causality are we to apply to the case where only an approximate measurement is effected (i.e., where the apparatus or observer interacts only weakly and for a finite time with the object-system)?

In the case of approximate measurement, we need to be supplied with rules which will tell us, for any initial object-system state, first, with what probability can we expect the various possible apparatus readings, and second, what new state to ascribe to the system after the value has been observed. We shall see that it is generally impossible to give these rules within a framework which considers the apparatus or observer as performing an (abstract) observation subject to Process 1, and that it is necessary, in order to give a full account of approximate measurements, to treat the entire system, including apparatus or observer, wave mechanically.

The position that an approximate measurement results in the situation that the object-system state is changed into an eigenstate of the exact measurement, but for which particular one the observer has only imprecise information, is manifestly false. It is a fact that we can make successive approximate position measurements of particles (in cloud chambers, for example) and use the results for somewhat reliable predictions of future positions. However, if either of these measurements left the particle in an "eigenstate" of position ( $\delta$ function), even though the particular one remained unknown, the momentum would have such a variance that no such prediction would be possible. (The possibility of such predictions lies in the correlations between position and momentum at one time with position and momentum at a later time for wave packets[9] – correlations which are totally destroyed by precise measurements of either quantity.)

Instead of continuing the discussion of the inadequacy of the probabilistic formulation, let us first investigate what actually happens in approximate measurements, from the viewpoint of pure wave mechanics. An approximate measurement consists of an interaction, for a finite time, which only imperfectly correlates the apparatus (or observer) with the object-system. We can deduce the desired rules in any particular case by the following method: For fixed interaction and initial apparatus state and for any initial object-system state we solve the wave equation for the time of interaction in question. The result will be a superposition of apparatus (observer) states and relative object-system states. Then (according to the method of Chapter IV for assigning a measure to a superposition) we assign a probability to each observed result equal to the square-amplitude of the coefficient of the element which contains the apparatus (observer) state representing the registering of that result. Finally, the object-system is assigned the new state which is its *relative state* in that element.

For example, let us consider the measuring process described in Chapter III - §3, which is an excellent model for an approximate measurement. After the interaction, the total state was found to be (III - (3.12)):

---

9 See Bohm [1], p. 202.

$$\psi_t^{S+A} = \int \frac{1}{N_{r'}} \xi^{r'}(q)\, \delta(r-r')\, dr' \; . \tag{4.1}$$

Then, according to our prescription, we assign the probability density $P(r')$ to the observation of the apparatus coordinate $r'$

$$P(r') = \left|\frac{1}{N_{r'}}\right|^2 = \int \phi^*\phi(q)\,\eta^*\eta\,(r'-qt)\,dq \; , \tag{4.2}$$

which is the square amplitude of the coefficient $\left(\frac{1}{N_{r'}}\right)$ of the element $\xi^{r'}(q)\delta(r-r')$ of the superposition (4.1) in which the apparatus coordinate has the value $r = r'$. Then, depending upon the observed apparatus coordinate $r'$, we assign the object-system the new state

$$\xi^{r'}(q) = N_{r'}\phi(q)\eta(r'-qt) \tag{4.3}$$

(where $\phi(q)$ is the old state, and $\eta(r)$ is the initial apparatus state) which is the relative object-system state in (4.1) for apparatus coordinate $r'$.

This example supplies the counter-example to another conceivable method of dealing with approximate measurement within the framework of Process 1. This is the position that when an approximate measurement of a quantity Q is performed, in actuality another quantity Q′ is precisely measured, where the eigenstates of Q′ correspond to fairly well-defined (i.e., sharply peaked distributions for) Q values.[10] However, any such scheme based on Process 1 always has the prescription that after the measurement, the (unnormalized) new state function results from the old by a projection (on an eigenstate or eigenspace), which depends upon the observed value. If this is true, then in the above example the new state $\xi^{r'}(q)$ must result from the old, $\phi(q)$, by a projection E:

$$\xi^{r'}(q) = N\,E\phi(q) = N_{r'}\phi(q)\eta(r'-qt) \tag{4.4}$$

[10] Cf. von Neumann [17], Chapter IV, §4.

where $N, N_{r'}$ are normalization constants). But $E$ is only a projection if $E^2 = E$. Applying the operation (4.4) twice, we get:

$$(4.5) \qquad E(N E\phi(q)) = N E^2\phi(q) = N'\phi(q)\eta^2(r'-qt) \Rightarrow E^2\phi(q) = \frac{N'}{N}\phi(q)\eta^2(r'-qt) \, ,$$

and we see that $E$ cannot be a projection unless $\eta(q) = \eta^2(q)$ for all $q$ (i.e., $\eta(q) = 0$ or $1$ for all $q$) and we have arrived at a contradiction to the assumption that in all cases the changes of states for approximate measurements are governed by projections. (In certain special cases, such as approximate position measurements with slits or Geiger counters,[11] the new functions arise from the old by multiplication by sharp cutoff functions which are 1 over the slit or counter and 0 elsewhere, so that these measurements *can* be handled by projections.)

One cannot, therefore, account for approximate measurements by any scheme based on Process 1, and it is necessary to investigate these processes entirely wave-mechanically. Our viewpoint constitutes a framework in which it is possible to make precise deductions about such measurements and observations, since we can follow in detail the interaction of an observer or apparatus with an object-system.

### §5. *Discussion of a spin measurement example*

We shall conclude this chapter with a discussion of an instructive example of Bohm.[12] Bohm considers the measurement of the $z$ component of the angular momentum of an atom, whose total angular momentum is $\frac{\hbar}{2}$, which is brought about by a Stern-Gerlach experiment. The measurement

[11] Cf. §2, this chapter.

[12] Bohm [1], p. 593.

is accomplished by passing an atomic beam through an inhomogeneous magnetic field, which has the effect of giving the particle a momentum which is directed up or down depending upon whether the spin was up or down.

The measurement is treated as impulsive, so that during the time that the atom passes through the field the Hamiltonian is taken to be simply the interaction:

$$H_I = \mu(\vec{\delta} \cdot \vec{\mathcal{H}}) \ , \quad \mu = -\frac{e\hbar}{2mc} \tag{5.1}$$

where $\mathcal{H}$ is the magnetic field and $\vec{\delta}$ the spin operator for the atom. The particle is presumed to pass through a region of the field where the field is in the $z$ direction, so that during the time of transit the field is approximately $\mathcal{H}_z \cong \mathcal{H}_0 + z\mathcal{H}_0' \left(\mathcal{H}_0 = (\mathcal{H}_z)_{z=0} \text{ and } \mathcal{H}_0' = \left(\frac{\partial \mathcal{H}_z}{\partial z}\right)_{z=0}\right)$, and hence the interaction is approximately:

$$H_I \cong \mu(\mathcal{H}_0 + z\mathcal{H}_0')S_z \ , \tag{5.2}$$

where $S_z$ denotes the operator for the $z$ component of the spin.

It is assumed that the state of the atom, just prior to entry into the field, is a wave packet of the form:

$$\psi_0 = f_0(z)(c_+ v_+ + c_- v_-) \tag{5.3}$$

where $v_+$ and $v_-$ are the spin functions for $S_z = 1$ and $-1$ respectively. Solving the Schrödinger equation for the Hamiltonian (5.2) and initial condition (5.3) yields the state for a later time $t$:

$$\psi = f_0(z)\left(c_+ e^{-i\mu(\mathcal{H}_0 + z\mathcal{H}_0')t/\hbar} v_+ + c_- e^{+i\mu(\mathcal{H}_0 + z\mathcal{H}_0')t/\hbar} v_-\right) . \tag{5.4}$$

Therefore, if $\Delta t$ is the time that it takes the atom to traverse the field,[13] each component of the wave packet has been multiplied by a phase factor

---

[13] This time is, strictly speaking, not well defined. The results, however, do not depend critically upon it.

$e^{\pm i\mu(\mathcal{H}_0 + z\mathcal{H}_0')\Delta t/\hbar}$, i.e., has had its mean momentum in the z direction changed by an amount $\pm\mathcal{H}_0'\mu\Delta t$, depending upon the spin direction. Thus the initial wave packet (with mean momentum zero) is split into a superposition of two packets, one with mean z-momentum $+\mathcal{H}_0'\mu\Delta t$ and spin up, and the other with spin down and mean z-momentum $-\mathcal{H}_0'\mu\Delta t$.

The interaction (5.2) has therefore served to correlate the spin with the momentum in the z-direction. These two packets of the resulting superposition now move in opposite z-directions, so that after a short time they become widely separated (provided that the momentum changes $\pm\mathcal{H}_0'\mu\Delta t$ are large compared to the momentum spread of the original packet), and the z-coordinate is itself then correlated with the spin — representing the "apparatus" coordinate in this case. The Stern-Gerlach apparatus therefore splits an incoming wave packet into a superposition of two diverging packets, corresponding to the two spin values.

We take this opportunity to caution against a certain viewpoint which can lead to difficulties. This is the idea that, after an apparatus has interacted with a system, in "actuality" one or another of the elements of the resultant superposition described by the composite state-function has been realized to the exclusion of the rest, the existing one simply being unknown to an external observer (i.e., that instead of the superposition there is a genuine mixture). This position must be erroneous since there is always the possibility for the external observer to make use of interference properties between the elements of the superposition.

In the present example, for instance, it is in principle possible to deflect the two beams back toward one another with magnetic fields and recombine them in another inhomogeneous field, which duplicates the first, in such a manner that the original spin state (before entering the apparatus) is restored.[14] This would not be possible if the original Stern-Gerlach apparatus performed the function of converting the original wave

14 As pointed out by Bohm [1], p. 604.

packet into a non-interfering mixture of packets for the two spin cases. Therefore the position that after the atom has passed through the inhomogeneous field it is "really" in one or the other beam with the corresponding spin, although we are ignorant of which one, is incorrect.

After two systems have interacted and become correlated it is true that marginal expectations for *subsystem* operators can be calculated correctly when the composite system is represented by a certain non-interfering mixture of states. Thus if the composite system state is $\psi^{S_1+S_2} = \sum_i a_i \phi_i^{S_1} \eta_i^{S_2}$, where the $\{\eta_i\}$ are orthogonal, then for purposes of calculating the expectations of operators on $S_1$ the state $\psi^{S_1+S_2}$ is equivalent to the non-interfering mixture of states $\phi_i^{S_1} \eta_i^{S_2}$ weighted by $P_i = a_i^* a_i$, and one can take the picture that one or another of the cases $\phi_i^{S_1} \eta_i^{S_2}$ has been realized to the exclusion of the rest, with probabilities $P_i$.[15]

However, this representation by a mixture must be regarded as only a mathematical artifice which, although useful in many cases, is an *incomplete description* because it ignores phase relations between the separate elements which actually exist, and which become important in any interactions which involve more than just a subsystem.

In the present example, the "composite system" is made of the "subsystems" spin value (object-system) and z-coordinate (apparatus), and the superposition of the two diverging wave packets is the state after interaction. It is only correct to regard this state as a mixture so long as any contemplated future interactions or measurements will involve only the spin value or only the z-coordinate, but not both simultaneously. As we saw, phase relations between the two packets are present and become important when they are deflected back and recombined in another inhomogeneous field – a process involving the spin values and z-coordinate simultaneously.

---

15 See Chapter III, §1.

It is therefore improper to attribute any less validity or "reality" to any element of a superposition than any other element, due to this ever present possibility of obtaining interference effects between the elements. All elements of a superposition must be regarded as simultaneously existing.

At this time we should like to add a few remarks concerning the notion of *transition probabilities* in quantum mechanics. Often one considers a system, with Hamiltonian $H$ and stationary states $\{\phi_i\}$, to be perturbed for a time by a time-dependent addition to the Hamiltonian, $H_I(t)$. Then under the action of the perturbed Hamiltonian $H' = H + H_I(t)$ the states $\{\phi_i\}$ are generally no longer stationary but change after time $t$ into new states $\{\psi_i(t)\}$:

(5.5) $$\phi_i \rightarrow \psi_i(t) = \sum_j (\phi_j, \psi_i(t))\phi_j = \sum_j a_{ij}(t)\phi_j ,$$

which can be represented as a superposition of the old stationary states with time-dependent coefficients $a_{ij}(t)$.

If at time $\tau$ a measurement with eigenstates $\phi_j$ is performed, such as an energy measurement (whose operator is the original $H$), then according to the probabilistic interpretation the probability for finding the state $\phi_j$, given that the state was originally $\phi_i$, is $P_{ij}(\tau) = |a_{ij}(\tau)|^2$. The quantities $|a_{ij}(\tau)|^2$ are often referred to as *transition probabilities*. In this case, however, the name is a misnomer, since it carries the connotation that the original state $\phi_i$ is transformed into a *mixture* (of the $\phi_j$ weighted by $P_{ij}(\tau)$), and gives the erroneous impression that the quantum formalism itself implies the existence of quantum-jumps (stochastic processes) independent of acts of observation. This is incorrect since there is still a pure state $\sum_j a_{ij}(\tau)\phi_j$ with phase relations between the $\phi_j$, and expectations of operators other than the energy *must* be calculated from the superposition and not the mixture.

There is another case, however, the one usually encountered in fact, where the transition probability concept is somewhat more justified. This is the case in which the perturbation is due to interaction of the system $s_1$ with another system $s_2$, and not simply a time dependence of $s_1$'s Hamiltonian as in the case just considered. In this situation the interaction produces a *composite system state*, for which there are in general no independent subsystem states. However, as we have seen, for purposes of calculating expectations of operators on $s_1$ alone, we can regard $s_1$ as being represented by a certain mixture. According to this picture the states of subsystem $s_1$ are gradually converted into mixtures by the interaction with $s_2$ and the concept of transition probability makes some sense. Of course, it must be remembered that this picture is only justified so long as further measurements on $s_1$ alone are contemplated, and any attempt to make a simultaneous determination in $s_1$ and $s_2$ involves the composite state where interference properties may be important.

An example is a hydrogen atom interacting with the electromagnetic field. After a time of interaction we can picture the atom as being in a mixture of its states, so long as we consider future measurements on the atom only. But in actuality the state of the atom is dependent upon (correlated with) the state of the field, and some process involving both atom and field could conceivably depend on interference effects between the states of the alleged mixture. With these restrictions, however, the concept of transition probability is quite useful and justified.

## VI. DISCUSSION

We have shown that our theory based on pure wave mechanics, which takes as the basic description of physical systems the state function – supposed to be an *objective* description (i.e., in one-one, rather than statistical, correspondence to the behavior of the system) – can be put in satisfactory correspondence with experience. We saw that the probabilistic assertions of the usual interpretation of quantum mechanics can be *deduced* from this theory, in a manner analogous to the methods of classical statistical mechanics, as subjective appearances to observers – observers which were regarded simply as physical systems subject to the same type of description and laws as any other systems, and having no preferred position. The theory is therefore capable of supplying us with a complete conceptual model of the universe, consistent with the assumption that it contains more than one observer.

Because the theory gives us an objective description, it constitutes a framework in which a number of puzzling subjects (such as classical level phenomena, the measuring process itself, the inter-relationship of several observers, questions of reversibility and irreversibility, etc.) can be investigated in detail in a logically consistent manner. It supplies a new way of viewing processes, which clarifies many apparent paradoxes of the usual interpretation[1] – indeed, it constitutes an objective framework in which it is possible to understand the general consistency of the ordinary view.

---

1 Such as that of Einstein, Rosen, and Podolsky [8], as well as the paradox of the introduction.

We shall now resume our discussion of alternative interpretations. There has been expressed lately a great deal of dissatisfaction with the present form of quantum theory by a number of authors, and a wide variety of new interpretations have sprung into existence. We shall now attempt to classify briefly a number of these interpretations, and comment upon them.

a. *The "popular" interpretation.* This is the scheme alluded to in the introduction, where $\psi$ is regarded as objectively characterizing the single system, obeying a deterministic wave equation when the system is isolated but changing probabilistically and discontinuously under observation.

In its unrestricted form this view can lead to paradoxes like that mentioned in the introduction, and is therefore untenable. However, this view *is* consistent so long as it is assumed that there is only one observer in the universe (the solipsist position – Alternative 1 of the Introduction). This consistency is most easily understood from the viewpoint of our own theory, where we were able to show that all phenomena will *seem* to follow the predictions of this scheme to any observer. Our theory therefore justifies the personal adoption of this probabilistic interpretation, for purposes of making practical predictions, from a more satisfactory framework.

b. *The Copenhagen interpretation.* This is the interpretation developed by Bohr. The $\psi$ function is not regarded as an objective description of a physical system (i.e., it is in no sense a conceptual model), but is regarded as merely a mathematical artifice which enables one to make statistical predictions, albeit the best predictions which it is possible to make. This interpretation in fact denies the very possibility of a single conceptual model applicable to the quantum realm, and asserts that the totality of phenomena can only be understood by the use of different, mutually exclusive (i.e., "complementary") models in different situations. All state-

> ments about microscopic phenomena are regarded as meaningless unless accompanied by a complete description (classical) of an experimental arrangement.

While undoubtedly safe from contradiction, due to its extreme conservatism, it is perhaps overcautious. We do not believe that the primary purpose of theoretical physics is to construct "safe" theories at severe cost in the applicability of their concepts, which is a sterile occupation, but to make useful models which serve for a time and are replaced as they are outworn.[2]

Another objectionable feature of this position is its strong reliance upon the classical level from the outset, which precludes any possibility of explaining this level on the basis of an underlying quantum theory. (The deduction of classical phenomena from quantum theory is impossible simply because no meaningful statements can be made without pre-existing classical apparatus to serve as a reference frame.) This interpretation suffers from the dualism of adhering to a "reality" concept (i.e., the possibility of objective description) on the classical level but renouncing the same in the quantum domain.

c. *The "hidden variables" interpretation.* This is the position (Alternative 4 of the Introduction) that $\psi$ is not a complete description of a single system. It is assumed that the correct complete description, which would involve further (hidden) parameters, would lead to a deterministic theory, from which the probabilistic aspects arise as a result of our ignorance of these extra parameters in the same manner as in classical statistical mechanics.

---

2 Cf. Appendix II.

The $\psi$-function is therefore regarded as a description of an *ensemble* of systems rather than a single system. Proponents of this interpretation include Einstein,[3] Bohm,[4] Wiener and Siegal.[5]

Einstein hopes that a theory along the lines of his general relativity, where all of physics is reduced to the geometry of space-time could satisfactorily explain quantum effects. In such a theory a particle is no longer a simple object but possesses an enormous amount of structure (i.e., it is thought of as a region of space-time of high curvature). It is conceivable that the interactions of such "particles" would depend in a sensitive way upon the details of this structure, which would then play the role of the "hidden variables."[6] However, these theories are non-linear and it is enormously difficult to obtain any conclusive results. Nevertheless, the possibility cannot be discounted.

Bohm considers $\psi$ to be a real force field acting on a particle which always has a well-defined position and momentum (which are the hidden variables of this theory). The $\psi$-field satisfying Schrödinger's equation is pictured as somewhat analogous to the electromagnetic field satisfying Maxwell's equations, although for systems of $n$ particles the $\psi$-field is in a $3n$-dimensional space. With this theory Bohm succeeds in showing that in all actual cases of measurement the best predictions that can be made are those of the usual theory, so that no experiments could ever rule out his interpretation in favor of the ordinary theory. Our main criticism of this view is on the grounds of simplicity – if one desires to hold the view that $\psi$ is a real field then the associated particle is superfluous since, as we have endeavored to illustrate, the pure wave theory is itself satisfactory.

---

3 Einstein [7].

4 Bohm [2].

5 Wiener and Siegal [20].

6 For an example of this type of theory see Einstein and Rosen [9].

Wiener and Siegal have developed a theory which is more closely tied to the formalism of quantum mechanics. From the set N of all non-degenerate linear Hermitian operators for a system having a complete set of eigenstates, a subset I is chosen such that no two members of I commute and every element outside I commutes with at least one element of I. The set I therefore contains precisely one operator for every orientation of the principal axes of the Hilbert space for the system. It is postulated that each of the operators of I corresponds to an independent observable which can take any of the real numerical values of the spectrum of the operator. This theory, in its present form, is a theory of infinitely[7] many "hidden variables," since a system is pictured as possessing (at each instant) a value for every one of these "observables" simultaneously, with the changes in these values obeying precise (deterministic) dynamical laws. However, the change of any one of these variables with time depends upon the entire set of observables, so that it is impossible ever to discover by measurement the complete set of values for a system (since only one "observable" at a time can be observed). Therefore, statistical ensembles are introduced, in which the values of all of the observables are related to points in a "differential space," which is a Hilbert space containing a measure for which each (differential space) coordinate has an independent normal distribution. It is then shown that the resulting statistical dynamics is in accord with the usual form of quantum theory.

It cannot be disputed that these theories are often appealing, and might conceivably become important should future discoveries indicate serious inadequacies in the present scheme (i.e., they might be more easily modified to encompass new experience). But from our viewpoint they are usually more cumbersome than the conceptually simpler theory based on pure wave mechanics. Nevertheless, these theories are of great theoretical importance because they provide us with examples that "hidden variables" theories are indeed possible.

---

[7] A non-denumerable infinity, in fact, since the set I is uncountable!

d. *The stochastic process interpretation.* This is the point of view which holds that the fundamental processes of nature are stochastic (i.e., probabilistic) processes. According to this picture physical systems are supposed to exist at all times in definite states, but the states are continually undergoing probabilistic changes. The discontinuous probabilistic "quantum-jumps" are not associated with acts of observation, but are fundamental to the systems themselves.

A stochastic theory which emphasizes the particle, rather than wave, aspects of quantum theory has been investigated by Bopp.[8] The particles do not obey deterministic laws of motion, but rather probabilistic laws, and by developing a general "correlation statistics" Bopp shows that his quantum scheme is a special case which gives results in accord with the usual theory. (This accord is only approximate and in principle one could decide between the theories. The approximation is so close, however, that it is hardly conceivable that a decision would be practically feasible.)

Bopp's theory seems to stem from a desire to have a theory founded upon particles rather than waves, since it is this particle aspect (highly localized phenomena) which is most frequently encountered in present day high-energy experiments (cloud chamber tracks, etc.). However, it seems to us to be much easier to understand particle aspects from a wave picture (concentrated wave packets) than it is to understand wave aspects (diffraction, interference, etc.) from a particle picture.

Nevertheless, there can be no fundamental objection to the idea of a stochastic theory, except on grounds of a naked prejudice for determinism. The question of determinism or indeterminism in nature is obviously forever undecidable in physics, since for any current deterministic [probabilistic] theory one could always postulate that a refinement of the theory

8 Bopp [5].

would disclose a probabilistic [deterministic] substructure, and that the current deterministic [probabilistic] theory is to be explained in terms of the refined theory on the basis of the law of large numbers [ignorance of hidden variables]. However, it is quite another matter to object to a mixture of the two where the probabilistic processes occur only with acts of observation.

e. *The wave interpretation.* This is the position proposed in the present thesis, in which the wave function itself is held to be the fundamental entity, obeying at all times a deterministic wave equation.

This view also corresponds most closely with that held by Schrödinger.[9] However, this picture only makes sense when observation processes themselves are treated within the theory. It is only in this manner that the *apparent* existence of definite macroscopic objects, as well as localized phenomena, such as tracks in cloud chambers, can be satisfactorily explained in a wave theory where the waves are continually diffusing. With the deduction in this theory that phenomena will appear to observers to be subject to Process 1, Heisenberg's criticism[10] of Schrödinger's opinion – that continuous wave mechanics could not seem to explain the discontinuities which are everywhere observed – is effectively met. The "quantum-jumps" exist in our theory as *relative* phenomena (i.e., the states of an object-system relative to chosen observer states show this effect), while the absolute states change quite continuously.

The wave theory is definitely tenable and forms, we believe, the simplest complete, self-consistent theory.

9 Schrodinger [18].

10 Heisenberg [14].

We should like now to comment on some views expressed by Einstein. Einstein's[11] criticism of quantum theory (which is actually directed more against what we have called the "popular" view than Bohr's interpretation) is mainly concerned with the drastic changes of state brought about by simple acts of observation (i.e., the infinitely rapid collapse of wave functions), particularly in connection with correlated systems which are widely separated so as to be mechanically uncoupled at the time of observation.[12] At another time he put his feeling colorfully by stating that he could not believe that a mouse could bring about drastic changes in the universe simply by looking at it.[13]

However, from the standpoint of our theory, *it is not so much the system which is affected by an observation as the observer, who becomes correlated to the system.*

In the case of observation of one system of a pair of spatially separated, correlated systems, nothing happens to the remote system to make any of its states more "real" than the rest. It had no independent states to begin with, but a number of states occurring in a superposition with corresponding states for the other (near) system. Observation of the near system simply correlates the observer to this system, a purely local process — but a process which also entails automatic correlation with the remote system. Each state of the remote system still exists with the same amplitude in a superposition, but now a superposition for which element contains, in addition to a remote system state and correlated near system state, an observer state which describes an observer who perceives the state of the near system.[14] From the present viewpoint all elements of

---

[11] Einstein [7].

[12] For example, the paradox of Einstein, Rosen, and Podolsky [8].

[13] Address delivered at Palmer Physical Laboratory, Princeton, Spring, 1954.

[14] See in this connection Chapter IV, particularly pp. 82, 83.

this superposition are equally "real." Only the observer state has changed, so as to become correlated with the state of the near system and hence naturally with that of the remote system also. The mouse does not affect the universe – only the mouse is affected.

Our theory in a certain sense bridges the positions of Einstein and Bohr, since the complete theory is quite objective and deterministic ("God does not play dice with the universe"), and yet on the subjective level, of assertions relative to observer states, it is probabilistic in the *strong sense* that there is no way for observers to make any predictions better than the limitations imposed by the uncertainty principle.[15]

In conclusion, we have seen that if we wish to adhere to objective descriptions then the principle of the psycho-physical parallelism requires that we should be able to consider some mechanical devices as representing observers. The situation is then that such devices must either cause the probabilistic discontinuities of Process 1, or must be transformed into the superpositions we have discussed. We are forced to abandon the former possibility since it leads to the situation that some physical systems would obey different laws from the rest, with no clear means for distinguishing between these two types of systems. We are thus led to our present theory which results from the complete abandonment of Process 1 as a basic process. Nevertheless, within the context of this theory, which is objectively deterministic, it develops that the probabilistic aspects of Process 1 reappear at the subjective level, as relative phenomena to observers.

One is thus free to build a conceptual model of the universe, which postulates only the existence of a universal wave function which obeys a linear wave equation. One then investigates the internal correlations in this wave function with the aim of deducing laws of physics, which are

---

[15] Cf. Chapter V, §2.

statements that take the form: Under the conditions C the property A of a subsystem of the universe (subset of the total collection of coordinates for the wave function) is correlated with the property B of another subsystem (with the manner of correlation being specified). For example, the classical mechanics of a system of massive particles becomes a law which expresses the correlation between the positions and momenta (approximate) of the particles at one time with those at another time.[16] All statements about subsystems then become *relative* statements, i.e., statements about the subsystem relative to a prescribed state for the remainder (since this is generally the only way a subsystem even possesses a unique state), and all laws are correlation laws.

The theory based on pure wave mechanics is a conceptually simple causal theory, which fully maintains the principle of the psycho-physical parallelism. It therefore forms a framework in which it is possible to discuss (in addition to ordinary phenomena) observation processes themselves, including the inter-relationships of several observers, in a logical, unambiguous fashion. In addition, all of the correlation paradoxes, like that of Einstein, Rosen, and Podolsky,[17] find easy explanation.

While our theory justifies the personal use of the probabilistic interpretation as an aid to making practical predictions, it forms a broader frame in which to understand the consistency of that interpretation. It transcends the probabilistic theory, however, in its ability to deal logically with questions of imperfect observation and approximate measurement.

Since this viewpoint will be applicable to all forms of quantum mechanics which maintain the superposition principle, it may prove a fruitful framework for the interpretation of new quantum formalisms. Field theories, particularly any which might be relativistic in the sense of general rela-

16 Cf. Chapter V, §2.

17 Einstein, Rosen, and Podolsky [8].

tivity, might benefit from this position, since one is free to construct formal (non-probabilistic) theories, and supply any possible statistical interpretations later. (This viewpoint avoids the necessity of considering anomalous probabilistic jumps scattered about space-time, and one can assert that field equations are satisfied everywhere and everywhen, then *deduce* any statistical assertions by the present method.)

By focusing attention upon questions of correlations, one may be able to deduce useful relations (correlation laws analogous to those of classical mechanics) for theories which at present do not possess known classical counterparts. Quantized fields do not generally possess pointwise independent field values, the values at one point of space-time being correlated with those at neighboring points of space-time in a manner, it is to be expected, approximating the behavior of their classical counterparts. If correlations are important in systems with only a finite number of degrees of freedom, how much more important they must be for systems of infinitely many coordinates.

Finally, aside from any possible practical advantages of the theory, it remains a matter of intellectual interest that the statistical assertions of the usual interpretation do not have the status of independent hypotheses, but are deducible (in the present sense) from the pure wave mechanics, which results from their omission.

## APPENDIX I

We shall now supply the proofs of a number of assertions which have been made in the text.

§1. *Proof of Theorem* 1

We now show that $\{X,Y,\ldots,Z\} > 0$ unless $X,Y,\ldots,Z$ are independent random variables. Abbreviate $P(x_i,y_j,\ldots,z_k)$ by $P_{ij\ldots k}$, and let

$$Q_{ij\ldots k} = \begin{cases} \dfrac{P_{ij\ldots k}}{P_iP_j\ldots P_k} & \text{if } P_iP_j\ldots P_k > 0 \\ 1 & \text{if } P_iP_j\ldots P_k = 0 \end{cases} \tag{1.1}$$

(Note that $P_iP_j\ldots P_k = 0$ implies that also $P_{ij\ldots k} = 0$.) Then always

$$P_{ij\ldots k} = Q_{ij\ldots k}P_iP_j\ldots P_k \ , \tag{1.2}$$

and we have

$$\{X,Y,\ldots,Z\} = \text{Exp}\left[\ln \frac{P_{ij\ldots k}}{P_iP_j\ldots P_k}\right] = \text{Exp}\ [\ \ln Q_{ij\ldots k}] \tag{1.3}$$

$$= \sum_{ij\ldots k} P_iP_j\ldots P_k\, Q_{ij\ldots k} \ln Q_{ij\ldots k}$$

Applying the inequality for $x \geqq 0$:

$$x \ln x > x - 1 \qquad \text{(except for } x = 1) \tag{1.4}$$

(which is easily established by calculating the minimum of $x \ln x - (x-1)$) to (1.3) we have:

$$P_iP_j\dots P_k\, Q_{ij\dots k} \ln Q_{ij\dots k} > P_iP_j\dots P_k (Q_{ij\dots k} - 1) \qquad (1.5)$$

$$(\text{unless } Q_{ij\dots k} = 1)\ .$$

Therefore we have for the sum:

$$\sum_{ij\dots k} P_iP_j\dots P_k Q_{ij\dots k} \ln Q_{ij\dots k} > \sum_{ij\dots k} P_iP_j\dots P_k Q_{ij\dots k} - \sum_{ij\dots k} P_iP_j\dots P_k\ , \qquad (1.6)$$

unless *all* $Q_{ij\dots k} = 1$. But $\sum_{ij\dots k} P_iP_j\dots P_k\, Q_{ij\dots k} = \sum_{ij\dots k} P_{ij\dots k} = 1$, and $\sum_{ij\dots k} P_iP_j\dots P_k = 1$, so that the right side of (1.6) vanishes. The left side is, by (1.3) the correlation $\{X,Y,\dots,Z\}$, and the condition that all of the $Q_{ij\dots k}$ equal one is precisely the independence condition that $P_{ij\dots k} = P_iP_j\dots P_k$ for all $i,j,\dots,k$. We have therefore proved that

$$\{X,Y,\dots,Z\} > 0 \qquad (1.7)$$

unless $X,Y,\dots,Z$ are mutually independent.

§2. *Convex function inequalities*

We shall now establish some basic inequalities which follow from the convexity of the function $x \ln x$.

LEMMA 1.
$$x_i \geqq 0,\quad P_i \geqq 0,\quad \sum_i P_i = 1$$
$$\Rightarrow \left(\sum_i P_ix_i\right) \ln \left(\sum_i P_ix_i\right) \leqq \sum_i P_ix_i \ln x_i\ .$$

This property is usually taken as the definition of a convex function,[1] but follows from the fact that the second derivative of $x \ln x$ is positive for all positive $x$, which is the elementary notion of convexity. There is also an immediate corollary for the continuous case:

1 See Hardy, Littlewood, and Pólya [13], p. 70.

COROLLARY 1. $\quad g(x) \geqq 0, \quad P(x) \geqq 0, \quad \int P(x)\,dx = 1$

$$\Rightarrow \left[\int P(x)\,g(x)\,dx\right] \ln \left[\int P(x)\,g(x)\,dx\right] \leqq \int P(x)\,g(x) \ln g(x)\,dx\,.$$

We can now derive a more general and very useful inequality from Lemma 1:

LEMMA 2. $\quad x_i \geqq 0, \quad a_i \geqq 0 \quad (\text{all } i)$

$$\Rightarrow \left(\sum_i x_i\right) \ln \left(\frac{\sum_i x_i}{\sum_i a_i}\right) \leqq \sum_i x_i \ln \left(\frac{x_i}{a_i}\right).$$

*Proof*: Let $P_i = a_i / \sum_i a_i$, so that $P_i \geqq 0$ and $\sum_i P_i = 1$. Then by Lemma 1:

$$\left[\sum_i P_i\left(\frac{x_i}{a_i}\right)\right] \ln \left[\sum_i P_i\left(\frac{x_i}{a_i}\right)\right] \leqq \sum_i P_i\left(\frac{x_i}{a_i}\right) \ln \left(\frac{x_i}{a_i}\right). \tag{2.1}$$

Substitution for $P_i$ yields:

$$\left[\sum_i \frac{a_i}{\left(\sum_i a_i\right)}\left(\frac{x_i}{a_i}\right)\right] \ln \left[\sum_i \frac{a_i}{\left(\sum_i a_i\right)}\left(\frac{x_i}{a_i}\right)\right] \leqq \sum_i \frac{a_i}{\left(\sum_i a_i\right)}\left(\frac{x_i}{a_i}\right) \ln \left(\frac{x_i}{a_i}\right), \tag{2.2}$$

which reduces to

$$\left(\sum_i x_i\right) \ln \left(\frac{\sum_i x_i}{\sum_i a_i}\right) \leqq \sum_i x_i \ln \left(\frac{x_i}{a_i}\right), \tag{2.3}$$

and we have proved the lemma.

We also mention the analogous result for the continuous case:

COROLLARY 2. $f(x) \geqq 0, \quad g(x) \geqq 0 \quad (\text{all } x)$

$$\Rightarrow \left[\int f(x)\,dx\right] \ln \left[\frac{\int f(x)\,dx}{\int g(x)\,dx}\right] \leqq \int f(x) \ln \left(\frac{f(x)}{g(x)}\right) dx \,.$$

§3. *Refinement theorems*

We now supply the proof for Theorems 2 and 4 of Chapter II, which concern the behavior of correlation and information upon refinement of the distributions. We suppose that the original (unrefined) distribution is $P_{ij\ldots k} = P(x_i, y_j, \ldots, z_k)$, and that the *refined* distribution is $P'^{\mu_i, \nu_j, \ldots, \eta_k}_{ij\ldots k}$, where the original value $x_i$ for $X$ has been resolved into a number of values $x_i^{\mu_i}$, and similarly for $Y, \ldots, Z$. Then:

$$(3.1) \qquad P_{ij\ldots k} = \sum_{\mu_i, \nu_j, \ldots, \eta_k} P'^{\mu_i, \nu_j, \ldots, \eta_k}_{ij\ldots k}, \quad P_i = \sum_{\mu_i} P'^{\mu_i}_i, \quad \text{etc.}$$

Computing the new correlation $\{X, Y, \ldots, Z\}'$ for the refined distribution $P'^{\mu_i, \nu_j, \ldots, \eta_k}_{ij\ldots k}$ we find:

$$(3.2) \quad \{X, Y, \ldots, Z\}' = \sum_{ij\ldots k} \sum_{\mu_i, \nu_j, \ldots, \eta_k} P'^{\mu_i, \nu_j, \ldots, \eta_k}_{ij\ldots k} \ln \left( \frac{P'^{\mu_i, \nu_j, \ldots, \eta_k}_{ij\ldots k}}{P'^{\mu_i}_i, P'^{\nu_j}_j, \ldots, P'^{\eta_k}_k} \right).$$

However, by Lemma 2, §2:

$$(3.3) \qquad \left( \sum_{\mu_i \ldots \eta_k} P'^{\mu_i \ldots \eta_k}_{i\ldots k} \right) \ln \left( \frac{\sum\limits_{\mu_i \ldots \eta_k} P'^{\mu_i \ldots \eta_k}_{i\ldots k}}{\sum\limits_{\mu_i \ldots \eta_k} P'^{\mu_i}_i, P'^{\nu_j}_j, \ldots, P'^{\eta_k}_k} \right)$$

$$\leqq \sum_{\mu_i \ldots \eta_k} P'^{\mu_i \ldots \eta_k}_{i\ldots k} \ln \left( \frac{P'^{\mu_i \ldots \eta_k}_{i\ldots k}}{P'^{\mu_i}_i, P'^{\nu_j}_j, \ldots, P'^{\eta_k}_k} \right).$$

Substitution of (3.3) into (3.2), noting that $\sum_{\mu_i \ldots \eta_k} P_i'^{\mu_i}, P_j'^{\nu_j}, \ldots, P_k'^{\eta_k}$ is equal to $\left(\sum_{\mu_i} P_i'^{\mu_i}\right)\left(\sum_{\nu_j} P_j'^{\nu_j}\right)\cdots\left(\sum_{\eta_k} P_k'^{\eta_k}\right)$, leads to:

(3.4)

$$\{X,Y,\ldots,Z\}' \geq \left(\sum_{ij\ldots k} \sum_{\mu_i \ldots \eta_k} P_{ij\ldots k}'^{\mu_i \ldots \eta_k}\right) \ln \left[\frac{\sum_{\mu_i \ldots \eta_k} P_{ij\ldots k}'^{\mu_i \ldots \eta_k}}{\left(\sum_{\mu_i} P_i'^{\mu_i}\right)\left(\sum_{\nu_j} P_j'^{\nu_j}\right)\cdots\left(\sum_{\eta_k} P_k'^{\eta_k}\right)}\right]$$

$$= \sum_{ij\ldots k} P_{ij\ldots k} \ln \frac{P_{ij\ldots k}}{P_i P_j \ldots P_k} = \{X,Y,\ldots,Z\} ,$$

and we have completed the proof of Theorem 2 (Chapter II), which asserts that refinement never decreases the correlation.[2]

We now consider the effect of refinement upon the relative information. We shall use the previous notation, and further assume that $a_i'^{\mu_i}, b_j'^{\nu_j}, \ldots, c_k'^{\eta_k}$ are the information measures for which we wish to compute the relative information of $P_{ij\ldots k}'^{\mu_i, \nu_j, \ldots, \eta_k}$ and of $P_{ij\ldots k}$. The information measures for the unrefined distribution $P_{ij\ldots k}$ then satisfy the relations:

$$a_i = \sum_{\mu_i} a_i'^{\mu_i} , \quad b_j = \sum_{\nu_j} b_j'^{\nu_j} , \ \ldots \ . \tag{3.5}$$

The relative information of the refined distribution is

$$I'_{XY\ldots Z} = \sum_{i\ldots j} \sum_{\mu_i \ldots \eta_k} P_{ij\ldots k}'^{\mu_i \ldots \eta_k} \ln \left[\frac{P_{ij\ldots k}'^{\mu_i \ldots \eta_k}}{a_i'^{\mu_i}, b_j'^{\nu_j}, \ldots, c_k'^{\eta_k}}\right] , \tag{3.6}$$

and by exactly the same procedure as we have just used for the correlation we arrive at the result:

---

2 Cf. Shannon [19], Appendix 7, where a quite similar theorem is proved.

$$I'_{XY\ldots Z} \geq \sum_{i\ldots k} P_{ij\ldots k} \ln \frac{P_{ij\ldots k}}{a_i b_j \ldots c_k} = I_{XY\ldots Z} \,, \tag{3.7}$$

and we have proved that refinement never decreases the relative information (Theorem 4, Chapter II).

It is interesting to note that the relation (3.4) for the behavior of correlation under refinement can be deduced from the behavior of relative information, (3.7). This deduction is an immediate consequence of the fact that the correlation is a relative information – the information of the *joint distribution* relative to the product measure of the *marginal distributions.*

§4. *Monotone decrease of information for stochastic processes*

We consider a sequence of transition-probability matrices $T^n_{ij}$ ($\sum_j T^n_{ij} = 1$ for all $n$, $i$, and $0 \leq T^n_{ij} \leq 1$ for all $n$, $i$, $j$), and a sequence of measures $a^n_i$ ($a^n_i \geq 0$) having the property that

$$a^{n+1}_j = \sum_i a^n_i T^n_{ij} \,. \tag{4.1}$$

We further suppose that we have a sequence of probability distributions, $P^n_i$, such that

$$P^{n+1}_j = \sum_i P^n_i T^n_{ij} \,. \tag{4.2}$$

For each of these probability distributions the relative information $I^n$ (relative to the $a^n_i$ measure) is defined:

$$I^n = \sum_i P^n_i \ln\left(\frac{P^n_i}{a^n_i}\right) . \tag{4.3}$$

Under these circumstances we have the following theorem:

THEOREM. $I^{n+1} \leq I^n$ .

*Proof*: Expanding $I^{n+1}$ we get:

$$(4.4) \qquad I^{n+1} = \sum_j P_j^{n+1} \ln \left(\frac{P_j^{n+1}}{a_j^{n+1}}\right) = \sum_j \left(\sum_i P_i^n T_{ij}^n\right) \ln \frac{\left(\sum_i P_i^n T_{ij}^n\right)}{\left(\sum_i a_i^n T_{ij}^n\right)} .$$

However, by Lemma 2 (§2, Appendix I) we have the inequality

$$(4.5) \qquad \left(\sum_i P_i^n T_{ij}^n\right) \ln \frac{\left(\sum_i P_i^n T_{ij}^n\right)}{\left(\sum_i a_i^n T_{ij}^n\right)} \leq \sum_i P_i^n T_{ij}^n \ln \frac{P_i^n T_{ij}^n}{a_i^n T_{ij}^n} .$$

Substitution of (4.5) into (4.4) yields:

$$(4.6) \qquad I^{n+1} \leq \sum_j \left(\sum_i P_i^n T_{ij}^n \ln \frac{P_i^n}{a_i^n}\right) = \sum_i P_i^n \left(\sum_j T_{ij}^n\right) \ln \left(\frac{P_i^n}{a_i^n}\right)$$

$$= \sum_i P_i^n \ln \left(\frac{P_i^n}{a_i^n}\right) = I^n ,$$

and the proof is completed.

This proof can be successively specialized to the case where $T$ is stationary ($T_{ij}^n = T_{ij}$ for all $n$) and then to the case where $T$ is doubly-stochastic ($\sum_i T_{ij} = 1$ for all $j$):

COROLLARY 1. $T_{ij}^n$ *is stationary* ($T_{ij}^n = T_{ij}$, *all* $n$), *and the measure* $a_i$ *is a stationary measure* ($a_j = \sum_i a_i T_{ij}$), *imply that the information,* $I^n = \sum_i P_i^n \ln (P_i^n/a_i^n)$, *is monotone decreasing. (As before,* $P_j^{n+1} = \sum_i P_i^n T_{ij}^n$.)

*Proof*: Immediate consequence of preceding theorem.

COROLLARY 2. $T_{ij}$ *is doubly-stochastic* $(\sum_i T_{ij} = 1$, *all* j) *implies that the information relative to the uniform measure* $(a_i = 1$, *all* i), $I^n = \sum_i P_i^n \ln P_i^n$, *is monotone decreasing.*

*Proof*: For $a_i = 1$ (all i) we have that $\sum_i a_i T_{ij} = \sum_i T_{ij} = 1 = a_j$. Therefore the uniform measure is stationary in this case and the result follows from Corollary 1.

These results hold for the continuous case also, and may be easily verified by replacing the above summations by integrations, and by replacing Lemma 2 by its corollary.

§5. *Proof of special inequality for Chapter IV* (1.7)

LEMMA. *Given probability densities* $P(r)$, $P_1(x)$, $P_2(r)$, *with* $P(r) = \int P_1(x) P_2(r - x\tau) dx$. *Then* $I_R \leqq I_X - \ln \tau$, *where* $I_X = \int P_1(x) \ln P_1(x) dx$ *and* $I_R = \int P(r) \ln P(r) dr$.

*Proof*: We first note that:

$$\int P_2(r - x\tau) dx = \int P_2(\omega) \frac{d\omega}{\tau} = \frac{1}{\tau} \qquad \text{(all } r) \tag{5.1}$$

and that furthermore

$$\int P_2(r - x\tau) dr = \int P_2(\omega) d\omega = 1 \qquad \text{(all } x) . \tag{5.2}$$

We now define the density $\widetilde{P}^r(x)$:

$$\widetilde{P}^r(x) = \tau P_2(r - x\tau) , \tag{5.3}$$

which is normalized, by (5.1). Then, according to §2, Corollary 1 Appendix I), we have the relation:

$$(5.4)\quad \left(\int \widetilde{P}^r(x)P_1(x)\,dx\right)\ln\left(\int \widetilde{P}^r(x)P_1(x)\,dx\right) \leqq \int \widetilde{P}^r(x)P_1(x)\,dx \ .$$

Substitution from (5.3) gives

$$(5.5)\quad \left(\tau\int P_2(r-x\tau)P_1(x)dx\right)\ln\left(\tau\int P_2(r-x\tau)P_1(x)dx\right)$$
$$\leqq \tau\int P_2(r-x\tau)P_1(x)\ln P_1(x)\,dx \ .$$

The relation $P(r) = \int P_1(x)P_2(r-x\tau)dx$, together with (5.5) then implies

$$(5.6)\quad P(r)\ln \tau P(r) \leqq \int P_2(r-x\tau)P_1(x)\ln P_1(x)\,dx \ ,$$

which is the same as:

$$(5.7)\quad P(r)\ln P(r) \leqq \int P_2(r-x\tau)P_1(x)\ln P_1(x)\,dx - P(r)\ln\tau \ .$$

Integrating with respect to r, and interchanging the order of integration on the right side gives:

$$(5.8)\quad I_R = \int P(r)\ln P(r)\,dr \leqq \int\left[\int P_2(r-x\tau)\,dr\right]P_1(x)\ln P_1(x)\,dx$$
$$-(\ln\tau)\int P(r)\,dr \ .$$

But using (5.2) and the fact that $\int P(r)\,dr = 1$ this means that

$$(5.9)\quad I_R \leqq \int P_1(x)\ln P_1(x)\,dx - \ln\tau = I_X - \ln\tau \ ,$$

and the proof of the lemma is completed.

§6. *Stationary point of* $I_K + I_X$

We shall show that the information sum:

$$(6.1)\quad I_K + I_X = \int_{-\infty}^{\infty} \phi^*\phi(k)\ln\phi^*\phi(k)\,dk + \int_{-\infty}^{\infty} \psi^*\psi(x)\ln\psi^*\psi(x)\,dx \ ,$$

where

$$\phi(k) = (1/\sqrt{2\pi})\int_{-\infty}^{\infty} e^{-ikx}\psi(x)\,dx$$

is *stationary* for the functions:

(6.2) $$\psi_0(x) = (1/2\pi\sigma_x^2)^{\frac{1}{4}}\, e^{-x^2/4\sigma^2_x}\,, \quad \phi_0(k) = (2\sigma_x^2/\pi)^{\frac{1}{4}}\, e^{-k^2\sigma^2_x}\,,$$

with respect to variations of $\psi$, $\delta\psi$, which preserve the normalization:

(6.3) $$\int_{-\infty}^{\infty} \delta(\psi^*\psi)\,dx = 0\,.$$

The variation $\delta\psi$ gives rise to a variation $\delta\phi$ of $\phi(k)$:

(6.4) $$\delta\phi = (1/\sqrt{2\pi})\int_{-\infty}^{\infty} e^{-ikx}\delta\psi\,dx\,.$$

To avoid duplication of effort we first calculate the variation $\delta I_\xi$ for an arbitrary wave function $u(\xi)$. By definition,

(6.5) $$I_\xi = \int_{-\infty}^{\infty} u^*(\xi)u(\xi)\ln u^*(\xi)u(\xi)\,d\xi\,,$$

so that

(6.6) $$\delta I_\xi = \int_{-\infty}^{\infty} [u^*u\,\delta(\ln u^*u) + \delta(u^*u)\ln u^*u]\,d\xi$$

$$= \int_{-\infty}^{\infty} (1+\ln u^*u)(u^*\delta u\;\; u\delta u^*)\,d\xi\,.$$

We now suppose that u has the *real* form:

(6.7) $$u(\xi) = a\,e^{-b\xi^2} = u^*(\xi)\,,$$

and from (6.6) we get

(6.8) $$\delta I_\xi = \int_{-\infty}^{\infty} (1+\ln a^2 - 2b\xi^2)\,a\,e^{-b\xi^2}(\delta u)\,d\xi + \text{complex conjugate.}$$

We now compute $\delta I_K$ for $\phi_0$ using (6.8), (6.2), and (6.4):

(6.9) $$\delta I_K\Big|_{\phi_0} = \int_{-\infty}^{\infty} (1+\ln a'^2 - 2b'k^2)a'e^{-b'k^2}\,\frac{1}{\sqrt{2\pi}}\int_{-\infty}^{\infty} e^{-ikx}\delta\psi\,dx\,dk + \text{c.c.}\,,$$

where

$$a = (2\sigma_x^2/\pi)^{\frac{1}{4}}, \quad b' = \sigma_x^2 .$$

Interchanging the order of integration and performing the definite integration over k we get:

$$(6.10) \qquad \delta I_K\Big|_{\phi_0} = \int_{-\infty}^{\infty} \frac{a'}{\sqrt{2b'}}\left(\ln a'^2 + \frac{x^2}{2b'}\right) e^{-(x^2/4b')}\delta\psi(x)\,dx + \text{c.c.} ,$$

while application of (6.8) to $\psi_0$ gives

$$(6.11) \qquad \delta I_X\Big|_{\psi_0} = \int_{-\infty}^{\infty} (1 + \ln a''^2 - 2b''x^2)\,a''e^{-b''x^2}\,\delta\psi(x)\,dx + \text{c.c.} ,$$

where

$$a'' = (1/2\pi\sigma_x^2)^{\frac{1}{4}}, \quad b'' = (1/4\sigma_x^2) .$$

Adding (6.10) and (6.11), and substituting for $a'$, $b'$, $a''$, $b''$, yields:

$$(6.12) \quad \delta(I_K+I_X)\Big|_{\psi_0} = (1-\ln\pi)\int_{-\infty}^{\infty} (1/2\pi\sigma_x^2)^{\frac{1}{4}}\, e^{-(x^2/4\sigma_x^2)}\,\delta\psi(x)\,dx + \text{c.c.}$$

But the integrand of (6.12) is simply $\psi_0(x)\delta\psi(x)$, so that

$$(6.13) \qquad \delta(I_K+I_X)\Big|_{\psi_0} = (1-\ln\pi)\int_{-\infty}^{\infty} \psi_0\,\delta\psi\,dx + \text{c.c.} .$$

Since $\psi_0$ is real, $\psi_0\delta\psi + \text{c.c.} = \psi_0^*\delta\psi + \text{c.c.} = \psi_0^*\delta\psi + \psi_0\delta\psi^* = \delta(\psi^*\psi)$, so that

$$(6.14) \qquad \delta(I_K+I_X)\Big|_{\psi_0} = (1-\ln\pi)\int_{-\infty}^{\infty} \delta(\psi^*\psi)\,dx = 0 ,$$

due to the normality restriction (6.3), and the proof is completed.

## APPENDIX II

## REMARKS ON THE ROLE OF THEORETICAL PHYSICS

There have been lately a number of new interpretations of quantum mechanics, most of which are equivalent in the sense that they predict the same results for all physical experiments. Since there is therefore no hope of deciding among them on the basis of physical experiments, we must turn elsewhere, and inquire into the fundamental question of the nature and purpose of physical theories in general. Only after we have investigated and come to some sort of agreement upon these general questions, i.e., of the role of theories themselves, will we be able to put these alternative interpretations in their proper perspective.

Every theory can be divided into two separate parts, the formal part, and the interpretive part. The formal part consists of a purely logico-mathematical structure, i.e., a collection of symbols together with rules for their manipulation, while the interpretive part consists of a set of "associations," which are rules which put some of the elements of the formal part into correspondence with the perceived world. The essential point of a theory, then, is that it is a *mathematical model*, together with an *isomorphism*[1] between the model and the world of experience (i.e., the sense perceptions of the individual, or the "real world" – depending upon one's choice of epistemology).

---

[1] By isomorphism we mean a mapping of some elements of the model into elements of the perceived world which has the property that the model is faithful, that is, if in the model a symbol A implies a symbol B, and A corresponds to the happening of an event in the perceived world, then the event corresponding to B must also obtain. The word homomorphism would be technically more correct, since there may not be a one-one correspondence between the model and the external world.

The model nature is quite apparent in the newest theories, as in nuclear physics, and particularly in those fields outside of physics proper, such as the Theory of Games, various economic models, etc., where the degree of applicability of the models is still a matter of considerable doubt. However, when a theory is highly successful and becomes firmly established, *the model tends to become identified with "reality" itself, and the model nature of the theory becomes obscured.* The rise of classical physics offers an excellent example of this process. The constructs of classical physics are just as much fictions of our own minds as those of any other theory; we simply have a great deal more confidence in them. It must be deemed a mistake, therefore, to attribute any more "reality" here than elsewhere.

Once we have granted that any physical theory is essentially only a model for the world of experience, we must renounce all hope of finding anything like "*the* correct theory." There is nothing which prevents any number of quite distinct models from being in correspondence with experience (i.e., all "correct"), and furthermore no way of ever verifying that any model is completely correct, simply because the totality of all experience is never accessible to us.

Two types of prediction can be distinguished; the prediction of phenomena already understood, in which the theory plays simply the role of a device for compactly summarizing known results (the aspect of most interest to the engineer), and the prediction of new phenomena and effects, unsuspected before the formulation of the theory. Our experience has shown that a theory often transcends the restricted field in which it was formulated. It is this phenomenon (which might be called the "inertia" of theories) which is of most interest to the theoretical physicist, and supplies a greater motive to theory construction than that of aiding the engineer.

From the viewpoint of the first type of prediction we would say that the "best" theory is the one from which the most accurate predictions can be most easily deduced — two not necessarily compatible ideals.

Classical physics, for example, permits deductions with far greater ease than the more accurate theories of relativity and quantum mechanics, and in such a case we must retain them all. It would be the worst sort of folly to advocate that the study of classical physics be completely dropped in favor of the newer theories. It can even happen that several quite distinct models can exist which are completely equivalent in their predictions, such that different ones are most applicable in different cases, a situation which seems to be realized in quantum mechanics today. It would seem foolish to attempt to reject all but one in such a situation, where it might be profitable to retain them all.

Nevertheless, we have a strong desire to construct a single all-embracing theory which would be applicable to the entire universe. From what stems this desire? The answer lies in the second type of prediction – the discovery of new phenomena – and involves the consideration of inductive inference and the factors which influence our *confidence* in a given theory (to be applicable outside of the field of its formulation). This is a difficult subject, and one which is only beginning to be studied seriously. Certain main points are clear, however, for example, that our confidence increases with the number of successes of a theory. If a new theory replaces several older theories which deal with separate phenomena, i.e., a comprehensive theory of the previously diverse fields, then our confidence in the new theory is very much greater than the confidence in either of the older theories, since the range of success of the new theory is much greater than any of the older ones. It is therefore this factor of confidence which seems to be at the root of the desire for comprehensive theories.

A closely related criterion is *simplicity* – by which we refer to conceptual simplicity rather than ease in use, which is of paramount interest to the engineer. A good example of the distinction is the theory of general relativity which is conceptually quite simple, while enormously cumbersome in actual calculations. Conceptual simplicity, like comprehensiveness, has the property of increasing confidence in a theory. A theory containing many *ad hoc* constants and restrictions, or many independent hypotheses, in no way impresses us as much as one which is largely free of arbitrariness.

It is necessary to say a few words about a view which is sometimes expressed, the idea that a physical theory should contain no elements which do not correspond directly to observables. This position seems to be founded on the notion that the only purpose of a theory is to serve as a summary of known data, and overlooks the second major purpose, the discovery of totally new phenomena. The major motivation of this viewpoint appears to be the desire to construct perfectly "safe" theories which will never be open to contradiction. Strict adherence to such a philosophy would probably seriously stifle the progress of physics.

The critical examination of just what quantities are observable in a theory does, however, play a useful role, since it gives an insight into ways of modification of a theory when it becomes necessary. A good example of this process is the development of Special Relativity. Such successes of the positivist viewpoint, when used merely as a tool for deciding which modifications of a theory are possible, in no way justify its universal adoption as a general principle which all theories must satisfy.

In summary, a physical theory is a logical construct (model), consisting of symbols and rules for their manipulation, some of whose elements are associated with elements of the perceived world. The fundamental requirements of a theory are logical consistency and correctness. There is no reason why there cannot be any number of different theories satisfying these requirements, and further criteria such as usefulness, simplicity, comprehensiveness, pictorability, etc., must be resorted to in such cases to further restrict the number. Even so, it may be impossible to give a total ordering of the theories according to "goodness," since different ones may rate highest according to the different criteria, and it may be most advantageous to retain more than one.

As a final note, we might comment upon the concept of *causality*. It should be clearly recognized that causality is a property of a model, and not a property of the world of experience. The concept of causality only makes sense with reference to a theory, in which there are logical dependences among the elements. A theory contains relations of the form "A implies B," which can be read as "A causes B," while our experience, uninterpreted by any theory, gives nothing of the sort, but only a *correlation* between the event corresponding to B and that corresponding to A.

## REFERENCES

[1] D. Bohm, *Quantum Theory*. Prentice-Hall, New York: 1951.

[2] D. Bohm, *Phys. Rev.* 84, 166, 1952 and 85, 180, 1952.

[3] N. Bohr, in *Albert Einstein, Philosopher–Scientist*. The Library of Living Philosophers, Inc., Vol. 7, p. 199. Evanston: 1949.

[4] N. Bohr, *Atomic Theory and the Description of Nature*.

[5] F. Bopp, *Z. Naturforsch.* 2a(4), 202, 1947; 7a 82, 1952; 8a, 6, 1953.

[6] J. L. Doob, *Stochastic Processes*. Wiley, New York: 1953.

[7] A. Einstein, in *Albert Einstein, Philosopher-Scientist*. The Library of Living Philosophers, Inc., Vol. 7, p. 665. Evanston: 1949.

[8] A. Einstein, B. Podolsky, N. Rosen, *Phys. Rev.* 47, 777, 1935.

[9] A. Einstein, N. Rosen, *Phys. Rev.* 48, 73, 1935.

[10] W. Feller, *An Introduction to Probability Theory and its Applications*. Wiley, New York: 1950.

[11] D. ter Haar, *Elements of Statistical Mechanics*. Rinehart, New York, 1954.

[12] P. R. Halmos, *Measure Theory*. Van Nostrand, New York: 1950.

[13] G. H. Hardy, J. E. Littlewood, G. Pólya, *Inequalities*. Cambridge University Press: 1952.

[14] W. Heisenberg, in *Niels Bohr and the Development of Physics*. McGraw-Hill, p. 12. New York: 1955.

[15] J. Kelley, *General Topology.* Van Nostrand, New York: 1955.

[16] A. I. Khinchin, *Mathematical Foundations of Statistical Mechanics.* (Translated by George Gamow) Dover, New York: 1949.

[17] J. von Neumann, *Mathematical Foundations of Quantum Mechanics.* (Translated by R. T. Beyer) Princeton University Press: 1955.

[18] E. Schrödinger, *Brit. J. Phil. Sci.* 3, 109, 233, 1952.

[19] C. E. Shannon, W. Weaver, *The Mathematical Theory of Communication.* University of Illinois Press: 1949.

[20] N. Wiener, I. E. Siegal, *Nuovo Cimento Suppl.* 2, 982 (1955).

[21] P. M. Woodward, *Probability and Information Theory, with Applications to Radar.* McGraw-Hill, New York: 1953.

Reprinted from REVIEWS OF MODERN PHYSICS, Vol. 29, No. 3, 454–462, July, 1957
Printed in U. S. A

# "Relative State" Formulation of Quantum Mechanics*

HUGH EVERETT, III†

*Palmer Physical Laboratory, Princeton University, Princeton, New Jersey*

## 1. INTRODUCTION

THE task of quantizing general relativity raises serious questions about the meaning of the present formulation and interpretation of quantum mechanics when applied to so fundamental a structure as the space-time geometry itself. This paper seeks to clarify the foundations of quantum mechanics. It presents a reformulation of quantum theory in a form believed suitable for application to general relativity.

The aim is not to deny or contradict the conventional formulation of quantum theory, which has demonstrated its usefulness in an overwhelming variety of problems, but rather to supply a new, more general and complete formulation, from which the conventional interpretation can be *deduced*.

The relationship of this new formulation to the older formulation is therefore that of a metatheory to a theory, that is, it is an underlying theory in which the nature and consistency, as well as the realm of applicability, of the older theory can be investigated and clarified.

The new theory is not based on any radical departure from the conventional one. The special postulates in the old theory which deal with observation are omitted in the new theory. The altered theory thereby acquires a new character. It has to be analyzed in and for itself before any identification becomes possible between the quantities of the theory and the properties of the world of experience. The identification, when made, leads back to the omitted postulates of the conventional theory that deal with observation, but in a manner which clarifies their role and logical position.

We begin with a brief discussion of the conventional formulation, and some of the reasons which motivate one to seek a modification.

## 2. REALM OF APPLICABILITY OF THE CONVENTIONAL OR "EXTERNAL OBSERVATION" FORMULATION OF QUANTUM MECHANICS

We take the conventional or "external observation" formulation of quantum mechanics to be essentially the following[1]: A physical system is completely described by a state function $\psi$, which is an element of a Hilbert space, and which furthermore gives information only to the extent of specifying the probabilities of the results of various observations which can be made *on* the system *by* external observers. There are two fundamentally different ways in which the state function can change:

*Process 1*: The discontinuous change brought about by the observation of a quantity with eigenstates $\phi_1, \phi_2, \cdots$, in which the state $\psi$ will be changed to the state $\phi_j$ with probability $|(\psi,\phi_j)|^2$.

*Process 2*: The continuous, deterministic change of state of an isolated system with time according to a wave equation $\partial\psi/\partial t = A\psi$, where $A$ is a linear operator.

This formulation describes a wealth of experience. No experimental evidence is known which contradicts it.

Not all conceivable situations fit the framework of this mathematical formulation. Consider for example an isolated system consisting of an observer or measuring apparatus, plus an object system. Can the change with time of the state of the *total* system be described by Process 2? If so, then it would appear that no discontinuous probabilistic process like Process 1 can take place. If not, we are forced to admit that systems which contain observers are not subject to the same kind of quantum-mechanical description as we admit for all other physical systems. The question cannot be ruled out as lying in the domain of psychology. Much of the discussion of "observers" in quantum mechanics has to do with photoelectric cells, photographic plates, and similar devices where a mechanistic attitude can hardly be contested. For the following one can *limit himself to this class of problems*, if he is unwilling to consider observers in the more familiar sense on the same mechanistic level of analysis.

What mixture of Processes 1 and 2 of the conventional formulation is to be applied to the case where only an approximate measurement is effected; that is, where an apparatus or observer interacts only weakly and for a limited time with an object system? In this case of an

---

* Thesis submitted to Princeton University March 1, 1957 in partial fulfillment of the requirements for the Ph.D. degree. An earlier draft dated January, 1956 was circulated to several physicists whose comments were helpful. Professor Niels Bohr, Dr. H. J. Groenewald, Dr. Aage Peterson, Dr. A. Stern, and Professor L. Rosenfeld are free of any responsibility, but they are warmly thanked for the useful objections that they raised. Most particular thanks are due to Professor John A. Wheeler for his continued guidance and encouragement. Appreciation is also expressed to the National Science Foundation for fellowship support.

† Present address: Weapons Systems Evaluation Group, The Pentagon, Washington, D. C.

[1] We use the terminology and notation of J. von Neumann, *Mathematical Foundations of Quantum Mechanics*, translated by R. T. Beyer (Princeton University Press, Princeton, 1955).

approximate measurement a proper theory must specify (1) the new state of the object system that corresponds to any particular reading of the apparatus and (2) the probability with which this reading will occur. von Neumann showed how to treat a special class of approximate measurements by the method of projection operators.[2] However, a general treatment of all approximate measurements by the method of projection operators can be shown (Sec. 4) to be impossible.

How is one to apply the conventional formulation of quantum mechanics to the space-time geometry itself? The issue becomes especially acute in the case of a closed universe.[3] There is no place to stand outside the system to observe it. There is nothing outside it to produce transitions from one state to another. Even the familiar concept of a proper state of the energy is completely inapplicable. In the derivation of the law of conservation of energy, one defines the total energy by way of an integral extended over a surface large enough to include all parts of the system and their interactions.[4] But in a closed space, when a surface is made to include more and more of the volume, it ultimately disappears into nothingness. Attempts to define a total energy for a closed space collapse to the vacuous statement, zero equals zero.

How are a quantum description of a closed universe, of approximate measurements, and of a system that contains an observer to be made? These three questions have one feature in common, that they all inquire about the *quantum mechanics* that is *internal to an isolated system*.

No way is evident to apply the conventional formulation of quantum mechanics to a system that is not subject to *external* observation. The whole interpretive scheme of that formalism rests upon the notion of external observation. The probabilities of the various possible outcomes of the observation are prescribed exclusively by Process 1. Without that part of the formalism there is no means whatever to ascribe a physical interpretation to the conventional machinery. But Process 1 is out of the question for systems not subject to external observation.[6]

### 3. QUANTUM MECHANICS INTERNAL TO AN ISOLATED SYSTEM

This paper proposes to regard pure wave mechanics (Process 2 only) as a complete theory. It postulates that a wave function that obeys a linear wave equation everywhere and at all times supplies a complete mathematical model for every isolated physical system without exception. It further postulates that every system that is subject to external observation can be regarded as part of a larger isolated system.

The wave function is taken as the basic physical entity with *no a priori interpretation*. Interpretation only comes *after* an investigation of the logical structure of the theory. Here as always the theory itself sets the framework for its interpretation.[5]

For any interpretation it is necessary to put the mathematical model of the theory into correspondence with experience. For this purpose it is necessary to formulate abstract models for observers that can be treated within the theory itself as physical systems, to consider isolated systems containing such model observers in interaction with other subsystems, to deduce the changes that occur in an observer as a consequence of interaction with the surrounding subsystems, and to interpret the changes in the familiar language of experience.

Section 4 investigates representations of the state of a composite system in terms of states of constituent subsystems. The mathematics leads one to recognize the concept of the *relativity of states*, in the following sense: a constituent subsystem cannot be said to be in any single well-defined state, independently of the remainder of the composite system. To any arbitrarily chosen state for one subsystem there will correspond a unique *relative state* for the remainder of the composite system. This relative state will usually depend upon the choice of state for the first subsystem. Thus the state of one subsystem does not have an independent existence, but is fixed only by the state of the remaining subsystem. In other words, the states occupied by the subsystems are not independent, but *correlated*. Such correlations between systems arise whenever systems interact. In the present formulation all measurements and observation processes are to be regarded simply as interactions between the physical systems involved—interactions which produce strong correlations. A simple model for a measurement, due to von Neumann, is analyzed from this viewpoint.

Section 5 gives an abstract treatment of the problem of observation. This uses only the superposition principle, and general rules by which composite system states are formed of subsystem states, in order that the results shall have the greatest generality and be applicable to any form of quantum theory for which these principles hold. Deductions are drawn about the state of the observer relative to the state of the object system. It is found that experiences of the observer (magnetic tape memory, counter system, etc.) are in full accord with predictions of the conventional "external observer" formulation of quantum mechanics, based on Process 1.

Section 6 recapitulates the "relative state" formulation of quantum mechanics.

---

[2] Reference 1, Chap. 4, Sec. 4.

[3] See A. Einstein, *The Meaning of Relativity* (Princeton University Press, Princeton, 1950), third edition, p. 107.

[4] L. Landau and E. Lifshitz, *The Classical Theory of Fields*, translated by M. Hamermesh (Addison-Wesley Press, Cambridge, 1951), p. 343.

[5] See in particular the discussion of this point by N. Bohr and L. Rosenfeld, Kgl. Danske Videnskab. Selskab, Mat.-fys. Medd. 12, No. 8 (1933).

#### 4. CONCEPT OF RELATIVE STATE

We now investigate some consequences of the wave mechanical formalism of composite systems. If a composite system $S$, is composed of two subsystems $S_1$ and $S_2$, with associated Hilbert spaces $H_1$ and $H_2$, then, according to the usual formalism of composite systems, the Hilbert space for $S$ is taken to be the *tensor product* of $H_1$ and $H_2$ (written $H=H_1\otimes H_2$). This has the consequence that if the sets $\{\xi_i^{S_1}\}$ and $\{\eta_j^{S_2}\}$ are complete orthonormal sets of states for $S_1$ and $S_2$, respectively, then the general state of $S$ can be written as a superposition:

$$\psi^S=\sum_{i,j}a_{ij}\xi_i^{S_1}\eta_j^{S_2}. \tag{1}$$

From (3.1) although $S$ is in a definite state $\psi^S$, the subsystems $S_1$ and $S_2$ do not possess anything like definite states independently of one another (except in the special case where all but one of the $a_{ij}$ are zero).

We *can*, however, for any choice of a state in one subsystem, *uniquely* assign a corresponding *relative* state in the other subsystem. For example, if we choose $\xi_k$ as the state for $S_1$, while the composite system $S$ is in the state $\psi^S$ given by (3.1), then the corresponding *relative state* in $S_2$, $\psi(S_2;\mathrm{rel}\,\xi_k,S_1)$, will be:

$$\psi(S_2;\mathrm{rel}\,\xi_k,S_1)=N_k\sum_j a_{kj}\eta_j^{S_2} \tag{2}$$

where $N_k$ is a normalization constant. This relative state for $\xi_k$ is *independent* of the choice of basis $\{\xi_i\}$ $(i\neq k)$ for the orthogonal complement of $\xi_k$, and is hence determined uniquely by $\xi_k$ alone. To find the relative state in $S_2$ for an arbitrary state of $S_1$ therefore, one simply carries out the above procedure using any pair of bases for $S_1$ and $S_2$ which contains the desired state as one element of the basis for $S_1$. To find states in $S_1$ relative to states in $S_2$, interchange $S_1$ and $S_2$ in the procedure.

In the conventional or "external observation" formulation, the relative state in $S_2$, $\psi(S_2;\mathrm{rel}\,\phi,S_1)$, for a state $\phi^{S_1}$ in $S_1$, gives the conditional probability distributions for the results of all measurements in $S_2$, given that $S_1$ has been measured and found to be in state $\phi^{S_1}$—i.e., that $\phi^{S_1}$ is the eigenfunction of the measurement in $S_1$ corresponding to the observed eigenvalue.

For any choice of basis in $S_1$, $\{\xi_i\}$, it is always possible to represent the state of $S$, (1), as a *single* superposition of pairs of states, each consisting of a state from the basis $\{\xi_i\}$ in $S_1$ and its relative state in $S_2$. Thus, from (2), (1) can be written in the form:

$$\psi^S=\sum_i\frac{1}{N_i}\xi_i^{S_1}\psi(S_2;\mathrm{rel}\,\xi_i,S_1). \tag{3}$$

This is an important representation used frequently.

*Summarizing: There does not, in general, exist anything like a single state for one subsystem of a composite system. Subsystems do not possess states that are independent of the states of the remainder of the system, so that the subsystem states are generally* correlated *with one another. One can arbitrarily choose a state for one subsystem, and be led to the relative state for the remainder. Thus we are faced with a fundamental* relativity of states, *which is implied by the formalism of composite systems. It is meaningless to ask the absolute state of a subsystem—one can only ask the state relative to a given state of the remainder of the subsystem.*

At this point we consider a simple example, due to von Neumann, which serves as a model of a measurement process. Discussion of this example prepares the ground for the analysis of "observation." We start with a system of only one coordinate, $q$ (such as position of a particle), and an apparatus of one coordinate $r$ (for example the position of a meter needle). Further suppose that they are initially independent, so that the combined wave function is $\psi_0^{S+A}=\phi(q)\eta(r)$ where $\phi(q)$ is the initial system wave function, and $\eta(r)$ is the initial apparatus function. The Hamiltonian is such that the two systems do not interact except during the interval $t=0$ to $t=T$, during which time the total Hamiltonian consists only of a simple interaction,

$$H_I=-i\hbar q(\partial/\partial r). \tag{4}$$

Then the state

$$\psi_t^{S+A}(q,r)=\phi(q)\eta(r-qt) \tag{5}$$

is a solution of the Schrödinger equation,

$$i\hbar(\partial\psi_t^{S+A}/\partial t)=H_I\psi_t^{S+A}, \tag{6}$$

for the specified initial conditions at time $t=0$.

From (5) at time $t=T$ (at which time interaction stops) there is no longer any definite independent apparatus state, nor any independent system state. The apparatus therefore does not indicate any definite object-system value, and nothing like process 1 has occurred.

Nevertheless, we *can* look upon the total wave function (5) as a *superposition* of pairs of subsystem states, each element of which has a definite $q$ value and a correspondingly displaced apparatus state. Thus after the interaction the state (5) has the form:

$$\psi_T^{S+A}=\int\phi(q')\delta(q-q')\eta(r-qT)dq', \tag{7}$$

which is a superposition of states $\psi_{q'}=\delta(q-q')\eta(r-qT)$. Each of these elements, $\psi_{q'}$, of the superposition describes a state in which the system has the definite value $q=q'$, and in which the apparatus has a state that is displaced from its original state by the amount $q'T$. These elements $\psi_{q'}$ are then superposed with coefficients $\phi(q')$ to form the total state (7).

Conversely, if we transform to the representation where the *apparatus* coordinate is definite, we write (5) as

$$\psi_T^{S+A}=\int(1/N_{r'})\xi^{r'}(q)\delta(r-r')dr',$$

where

$$\xi^{r'}(q) = N_{r'}\phi(q)\eta(r'-qT) \tag{8}$$

and

$$(1/N_{r'})^2 = \int \phi^*(q)\phi(q)\eta^*(r'-qT)\eta(r'-qT)dq.$$

Then the $\xi^{r'}(q)$ are the relative system state functions[6] for the apparatus states $\delta(r-r')$ of definite value $r=r'$.

If $T$ is sufficiently large, or $\eta(r)$ sufficiently sharp (near $\delta(r)$), then $\xi^{r'}(q)$ is nearly $\delta(q-r'/T)$ and the relative system states $\xi^{r'}(q)$ are nearly eigenstates for the values $q=r'/T$.

We have seen that (8) is a superposition of states $\psi_{r'}$, *for each of which* the apparatus has recorded a definite value $r'$, and the system is left in approximately the eigenstate of the measurement corresponding to $q=r'/T$. The discontinuous "jump" into an eigenstate is thus only a relative proposition, dependent upon the mode of decomposition of the total wave function into the superposition, and relative to a particularly chosen apparatus-coordinate value. So far as the complete theory is concerned all elements of the superposition exist simultaneously, and the entire process is quite continuous.

von Neumann's example is only a special case of a more general situation. Consider any measuring apparatus interacting with any object system. As a result of the interaction the state of the measuring apparatus is no longer capable of independent definition. It can be defined only *relative* to the state of the object system. In other words, there exists only a correlation between the states of the two systems. It seems as if nothing can ever be settled by such a measurement.

This indefinite behavior seems to be quite at variance with our observations, since physical objects always appear to us to have definite positions. Can we reconcile this feature wave mechanical theory built purely on Process 2 with experience, or must the theory be abandoned as untenable? In order to answer this question we consider the problem of observation itself within the framework of the theory.

## 5. OBSERVATION

We have the task of making deductions about the appearance of phenomena to observers which are considered as purely physical systems and are treated within the theory. To accomplish this it is necessary to identify some present properties of such an observer with features of the past experience of the observer. Thus, in order to say that an observer 0 has observed the event $\alpha$, it is necessary that the state of 0 has become changed from its former state to a new state which is dependent upon $\alpha$.

It will suffice for our purposes to consider the observers to possess memories (i.e., parts of a relatively permanent nature whose states are in correspondence with past experience of the observers). In order to make deductions about the past experience of an observer it is sufficient to deduce the present contents of the memory as it appears within the mathematical model.

As models for observers we can, if we wish, consider automatically functioning machines, possessing sensory apparatus and coupled to recording devices capable of registering past sensory data and machine configurations. We can further suppose that the machine is so constructed that its present actions shall be determined not only by its present sensory data, but by the contents of its memory as well. Such a machine will then be capable of performing a sequence of observations (measurements), and furthermore of deciding upon its future experiments on the basis of past results. If we consider that current sensory data, as well as machine configuration, is immediately recorded in the memory, then the actions of the machine at a given instant can be regarded as a function of the memory contents only, and all relavant experience of the machine is contained in the memory.

For such machines we are justified in using such phrases as "the machine has perceived $A$" or "the machine is aware of $A$" if the occurrence of $A$ is represented in the memory, since the future behavior of the machine will be based upon the occurrence of $A$. In fact, all of the customary language of subjective experience is quite applicable to such machines, and forms the most natural and useful mode of expression when dealing with their behavior, as is well known to individuals who work with complex automata.

When dealing with a system representing an observer quantum mechanically we ascribe a state function, $\psi^0$, to it. When the state $\psi^0$ describes an observer whose memory contains representations of the events $A, B, \cdots, C$ we denote this fact by appending the memory sequence in brackets as a subscript, writing:

$$\psi^0_{[A,B,\cdots,C]}. \tag{9}$$

The symbols $A, B, \cdots, C$, which we assume to be ordered time-wise, therefore stand for memory configurations which are in correspondence with the past experience of the observer. These configurations can be regarded as punches in a paper tape, impressions on a magnetic reel, configurations of a relay switching circuit, or even configurations of brain cells. We require only that they be capable of the interpretation "The observer has experienced the succession of events $A, B, \cdots, C$." (We sometimes write dots in a memory sequence, $\cdots A, B, \cdots, C$, to indicate the possible presence of previous

[6] This example provides a model of an approximate measurement. However, the relative system states after the interaction $\xi^{r'}(q)$ cannot ordinarily be generated from the original system state $\phi$ by the application of *any* projection operator, $E$. Proof: Suppose on the contrary that $\xi^{r'}(q) = NE\phi(q) = N'\phi(q)\eta(r'-qt)$, where $N$, $N'$ are normalization constants. Then

$$E(NE\phi(q)) = NE^2\phi(q) = N''\phi(q)\eta^2(r'-qt)$$

and $E^2\phi(q) = (N''/N)\phi(q)\eta^2(r'-qt)$. But the condition $E^2=E$ which is necessary for $E$ to be a projection implies that $N'/N''\ \eta(q) = \eta^2(q)$ which is generally false.

memories which are irrelevant to the case being considered.)

The mathematical model seeks to treat the interaction of such observer systems with other physical systems (observations), within the framework of Process 2 wave mechanics, and to deduce the resulting memory configurations, which are then to be interpreted as records of the past experiences of the observers.

We begin by defining what constitutes a "good" observation. A good observation of a quantity $A$, with eigenfunctions $\phi_i$, for a system $S$, by an observer whose initial state is $\psi^O$, consists of an interaction which, in a specified period of time, transforms each (total) state

$$\psi^{S+O}=\phi_i\psi^O[\ldots] \tag{10}$$

into a new state

$$\psi^{S+O'}=\phi_i\psi^O[\ldots\alpha_i] \tag{11}$$

where $\alpha_i$ characterizes[7] the state $\phi_i$. (The symbol, $\alpha_i$, might stand for a recording of the eigenvalue, for example.) That is, we require that the system state, *if it is an eigenstate*, shall be unchanged, and (2) that the observer state shall change so as to describe an observer that is "aware" of which eigenfunction it is; that is, some property is recorded in the memory of the observer which characterizes $\phi_i$, such as the eigenvalue. The requirement that the eigenstates for the system be unchanged is necessary if the observation is to be significant (repeatable), and the requirement that the observer state change in a manner which is different for each eigenfunction is necessary if we are to be able to call the interaction an observation at all. How closely a general interaction satisfies the definition of a good observation depends upon (1) the way in which the interaction depends upon the dynamical variables of the observer system—including memory variables—and upon the dynamical variables of the object system and (2) the initial state of the observer system. Given (1) and (2), one can for example solve the wave equation, deduce the state of the composite system after the end of the interaction, and check whether an object system that was originally in an eigenstate is left in an eigenstate, as demanded by the repeatability postulate. This postulate is satisfied, for example, by the model of von Neumann that has already been discussed.

From the definition of a good observation we first deduce the result of an observation upon a system which is *not* in an eigenstate of the observation. We know from our definition that the interaction transforms states $\phi_i\psi^O[\cdots]$ into states $\phi_i\psi^O[\cdots\alpha_i]$. Consequently these solutions of the wave equation can be superposed to give the final state for the case of an arbitrary initial system state. Thus if the initial system state is not an eigenstate, but a general state $\sum_i a_i\phi_i$, the final total state will have the form:

$$\psi^{S+O'}=\sum_i a_i\phi_i\psi^O[\cdots\alpha_i]. \tag{12}$$

This superposition principle continues to apply in the presence of further systems which do not interact during the measurement. Thus, if systems $S_1, S_2, \cdots, S_n$ are present as well as $O$, with original states $\psi^{S_1}, \psi^{S_2}, \cdots, \psi^{S_n}$, and the only interaction during the time of measurement takes place between $S_1$ and $O$, the measurement will transform the initial total state:

$$\psi^{S_1+S_2+\cdots+S_n+O}=\psi^{S_1}\psi^{S_2}\cdots\psi^{S_n}\psi^O[\cdots] \tag{13}$$

into the final state:

$$\psi'^{S_1+S_2+\cdots+S_n+O}=\sum_i a_i\phi_i^{S_1}\psi^{S_2}\cdots\psi^{S_n}\psi^O[\cdots\alpha_i] \tag{14}$$

where $a_i=(\phi_i^{S_1},\psi^{S_1})$ and $\phi_i^{S_1}$ are eigenfunctions of the observation.

Thus we arrive at the general rule for the transformation of total state functions which describe systems within which observation processes occur:

*Rule 1*: The observation of a quantity $A$, with eigenfunctions $\phi_i^{S_1}$, in a system $S_1$ by the observer $O$, transforms the total state according to:

$$\psi^{S_1}\psi^{S_2}\cdots\psi^{S_n}\psi^O[\cdots] \rightarrow \sum_i a_i\phi_i^{S_1}\psi^{S_2}\cdots\psi^{S_n}\psi^O[\cdots\alpha_i] \tag{15}$$

where

$$a_i=(\phi_i^{S_1},\psi^{S_1}).$$

If we next consider a *second* observation to be made, where our total state is now a superposition, we can apply Rule 1 separately to each element of the superposition, since each element separately obeys the wave equation and behaves independently of the remaining elements, and then superpose the results to obtain the final solution. We formulate this as:

*Rule 2*: Rule 1 may be applied separately to each element of a superposition of total system states, the results being superposed to obtain the final total state. Thus, a determination of $B$, with eigenfunctions $\eta_j^{S_2}$, on $S_2$ by the observer $O$ transforms the total state

$$\sum_i a_i\phi_i^{S_1}\psi^{S_2}\cdots\psi^{S_n}\psi^O[\cdots\alpha_i] \tag{16}$$

into the state

$$\sum_{i,j} a_i b_j\phi_i^{S_1}\eta_j^{S_2}\psi^{S_3}\cdots\psi^{S_n}\psi^O[\cdots\alpha_i,\beta_j] \tag{17}$$

where $b_j=(\eta_j^{S_2},\psi^{S_2})$, which follows from the application of Rule 1 to each element $\phi_i^{S_1}\psi^{S_2}\cdots\psi^{S_n}\psi^O[\cdots\alpha_i]$, and then superposing the results with the coefficients $a_i$.

These two rules, which follow directly from the superposition principle, give a convenient method for determining final total states for any number of observation processes in any combinations. We now seek the *interpretation* of such final total states.

[7] It should be understood that $\psi^O[\cdots\alpha_i]$ is a *different* state for each $i$. A more precise notation would write $\psi_i^O[\cdots\alpha_i]$, but no confusion can arise if we simply let the $\psi_i^O$ be indexed only by the index of the memory configuration symbol.

Let us consider the simple case of a single observation of a quantity $A$, with eigenfunctions $\phi_i$, in the system $S$ with initial state $\psi^S$, by an observer 0 whose initial state is $\psi^O[\cdots]$. The final result is, as we have seen, the superposition

$$\psi'^{S+O}=\sum_i a_i\phi_i\psi^O[\cdots\alpha_i]. \tag{18}$$

There is no longer any independent system state or observer state, although the two have become correlated in a one-one manner. However, in each *element* of the superposition, $\phi_i\psi^O[\cdots\alpha_i]$, the object-system state is a particular eigenstate of the observation, and *furthermore the observer-system state describes the observer as definitely perceiving that particular system state.* This correlation is what allows one to maintain the interpretation that a measurement has been performed.

We now consider a situation where the observer system comes into interaction with the object system for a second time. According to Rule 2 we arrive at the total state after the second observation:

$$\psi''^{S+O}=\sum_i a_i\phi_i\psi^O[\cdots\alpha_i,\alpha_i]. \tag{19}$$

Again, each element $\phi_i\psi^O[\cdots\alpha_i,\alpha_i]$ describes a system eigenstate, but this time also describes the observer as having obtained the *same result* for each of the two observations. Thus for every separate state of the observer in the final superposition the result of the observation was repeatable, even though different for different states. This repeatability is a consequence of the fact that after an observation the *relative* system state for a particular observer state is the corresponding eigenstate.

Consider now a different situation. An observer-system 0, with initial state $\psi^O[\cdots]$, measures the *same* quantity $A$ in a number of separate, identical, systems which are initially in the same state, $\psi^{S_1}=\psi^{S_2}=\cdots=\psi^{S_n}=\sum_i a_i\phi_i$ (where the $\phi_i$ are, as usual, eigenfunctions of $A$). The initial total state function is then

$$\psi_0{}^{S_1+S_2+\cdots+S_n+O}=\psi^{S_1}\psi^{S_2}\cdots\psi^{S_n}\psi^O[\cdots]. \tag{20}$$

We assume that the measurements are performed on the systems in the order $S_1, S_2, \cdots S_n$. Then the total state after the first measurement is by Rule 1,

$$\psi_1{}^{S_1+S_2+\cdots+S_n+O}=\sum_i a_i\phi_i{}^{S_1}\psi^{S_2}\cdots\psi^{S_n}\psi^O[\cdots\alpha_i{}^1] \tag{21}$$

(where $\alpha_i{}^1$ refers to the first system, $S_1$).

After the second measurement it is, by Rule 2,

$$\psi_2{}^{S_1+S_2+\cdots+S_n+O}=\sum_{i,j}a_ia_j\phi_i{}^{S_1}\phi_j{}^{S_2}\psi^{S_3}\cdots\psi^{S_n}\psi^O[\cdots\alpha_i{}^1,\alpha_j{}^2] \tag{22}$$

and in general, after $r$ measurements have taken place ($r\leq n$), Rule 2 gives the result:

$$\psi_r=\sum_{i,j,\cdots k}a_ia_j\cdots a_k\phi_i{}^{S_1}\phi_j{}^{S_2}\cdots\phi_k{}^{S_r}\psi^{S_{r+1}}\cdots\psi^{S_n}\psi^O[\cdots\alpha_i{}^1,\alpha_j{}^2,\cdots\alpha_k{}^r] \tag{23}$$

We can give this state, $\psi_r$, the following interpretation. It consists of a superposition of states:

$$\psi'_{ij\cdots k}=\phi_i{}^{S_1}\phi_j{}^{S_2}\cdots\phi_k{}^{S_r}\times\psi^{S_{r+1}}\cdots\psi^{S_n}\psi^O[\alpha_i{}^1,\alpha_j{}^2,\cdots\alpha_k{}^r] \tag{24}$$

each of which describes the observer with a definite memory sequence $[\alpha_i{}^1,\alpha_j{}^2,\cdots\alpha_k{}^r]$. Relative to him the (observed) system states are the corresponding eigenfunctions $\phi_i{}^{S_1},\phi_j{}^{S_2},\cdots,\phi_k{}^{S_r}$, the remaining systems, $S_{r+1},\cdots,S_n$, being unaltered.

A typical element $\psi'_{ij\cdots k}$ of the final superposition describes a state of affairs wherein the observer has perceived an apparently random sequence of definite results for the observations. Furthermore the object systems have been left in the corresponding eigenstates of the observation. At this stage suppose that a redetermination of an earlier system observation ($S_l$) takes place. Then it follows that every element of the resulting final superposition will describe the observer with a memory configuration of the form $[\alpha_i{}^1,\cdots\alpha_j{}^l,\cdots\alpha_k{}^r,\alpha_j{}^l]$ in which the earlier memory coincides with the later—i.e., the memory states are *correlated.* It will thus *appear* to the observer, as described by a typical element of the superposition, that each initial observation on a system caused the system to "jump" into an eigenstate in a random fashion and thereafter remain there for subsequent measurements on the same system. Therefore—disregarding for the moment quantitative questions of relative frequencies—the probabilistic assertions of Process 1 *appear* to be valid to the observer described by a typical element of the final superposition.

We thus arrive at the following picture: Throughout all of a sequence of observation processes there is only one physical system representing the observer, yet there is no single unique *state* of the observer (which follows from the representations of interacting systems). Nevertheless, there is a representation in terms of a *superposition*, each element of which contains a definite observer state and a corresponding system state. Thus with each succeeding observation (or interaction), the observer state "branches" into a number of different states. Each branch represents a different outcome of the measurement and the *corresponding* eigenstate for the object-system state. All branches exist simultaneously in the superposition after any given sequence of observations.‡

‡ *Note added in proof.*—In reply to a preprint of this article some correspondents have raised the question of the "transition from possible to actual," arguing that in "reality" there is—as our experience testifies—no such splitting of observer states, so that only one branch can ever actually exist. Since this point may occur to other readers the following is offered in explanation.

The whole issue of the transition from "possible" to "actual" is taken care of in the theory in a very simple way—there is no such transition, nor is such a transition necessary for the theory to be in accord with our experience. From the viewpoint of the theory *all* elements of a superposition (all "branches") are "actual," none any more "real" than the rest. It is unnecessary to suppose that all but one are somehow destroyed, since all the

The "trajectory" of the memory configuration of an observer performing a sequence of measurements is thus not a linear sequence of memory configurations, but a branching tree, with all possible outcomes existing simultaneously in a final superposition with various coefficients in the mathematical model. In any familiar memory device the branching does not continue indefinitely, but must stop at a point limited by the capacity of the memory.

In order to establish quantitative results, we must put some sort of measure (weighting) on the elements of a final superposition. This is necessary to be able to make assertions which hold for almost all of the observer states described by elements of a superposition. We wish to make quantitative statements about the relative frequencies of the different possible results of observation—which are recorded in the memory—for a typical observer state; but to accomplish this we must have a method for selecting a typical element from a superposition of orthogonal states.

We therefore seek a general scheme to assign a measure to the elements of a superposition of orthogonal states $\sum_i a_i\phi_i$. We require a positive function $m$ of the complex coefficients of the elements of the superposition, so that $m(a_i)$ shall be the measure assigned to the element $\phi_i$. In order that this general scheme be unambiguous we must first require that the states themselves always be normalized, so that we can distinguish the coefficients from the states. However, we can still only determine the *coefficients*, in distinction to the states, up to an arbitrary phase factor. In order to avoid ambiguities the function $m$ must therefore be a function of the amplitudes of the coefficients alone, $m(a_i)=m(|a_i|)$.

We now impose an additivity requirement. We can regard a subset of the superposition, say $\sum_{i=1}^{n} a_i\phi_i$, as a single element $\alpha\phi'$:

$$\alpha\phi'=\sum_{i=1}^{n} a_i\phi_i. \tag{25}$$

We then demand that the measure assigned to $\phi'$ shall be the sum of the measures assigned to the $\phi_i$ ($i$ from 1 to $n$):

$$m(\alpha)=\sum_{i=1}^{n} m(a_i). \tag{26}$$

Then we have already restricted the choice of $m$ to the square amplitude alone; in other words, we have $m(a_i)=a_i^*a_i$, apart from a multiplicative constant.

To see this, note that the normality of $\phi'$ requires that $|\alpha|=(\sum a_i^*a_i)^{\frac{1}{2}}$. From our remarks about the dependence of $m$ upon the amplitude alone, we replace the $a_i$ by their amplitudes $u_i=|a_i|$. Equation (26) then imposes the requirement,

$$m(\alpha)=m(\textstyle\sum a_i^*a_i)^{\frac{1}{2}}=m(\sum u_i^2)^{\frac{1}{2}} \\ =\textstyle\sum m(u_i)=\sum m(u_i^2)^{\frac{1}{2}}. \tag{27}$$

Defining a new function $g(x)$

$$g(x)=m(\sqrt{x}) \tag{28}$$

we see that (27) requires that

$$g(\textstyle\sum u_i^2)=\sum g(u_i^2). \tag{29}$$

Thus $g$ is restricted to be linear and necessarily has the form:

$$g(x)=cx \quad (c \text{ constant}). \tag{30}$$

Therefore $g(x^2)=cx^2=m(\sqrt{x^2})=m(x)$ and we have deduced that $m$ is restricted to the form

$$m(a_i)=m(u_i)=cu_i^2=ca_i^*a_i. \tag{31}$$

We have thus shown that the only choice of measure consistent with our additivity requirement is the square amplitude measure, apart from an arbitrary multiplicative constant which may be fixed, if desired, by normalization requirements. (The requirement that the total measure be unity implies that this constant is 1.)

The situation here is fully analogous to that of classical statistical mechanics, where one puts a measure on trajectories of systems in the phase space by placing a measure on the phase space itself, and then making assertions (such as ergodicity, quasi-ergodicity, etc.) which hold for "almost all" trajectories. This notion of "almost all" depends here also upon the choice of measure, which is in this case taken to be the Lebesgue measure on the phase space. One could contradict the statements of classical statistical mechanics by choosing a measure for which only the exceptional trajectories had nonzero measure. Nevertheless the choice of Lebesgue measure on the phase space can be justified by the fact that it is the only choice for which the "conservation of probability" holds, (Liouville's theorem) and hence the only choice which makes possible any reasonable statistical deductions at all.

In our case, we wish to make statements about "trajectories" of observers. However, for us a trajectory is constantly branching (transforming from state to superposition) with each successive measurement. To

---

separate elements of a superposition individually obey the wave equation with complete indifference to the presence or absence ("actuality" or not) of any other elements. This total lack of effect of one branch on another also implies that no observer will ever be aware of any "splitting" process.

Arguments that the world picture presented by this theory is contradicted by experience, because we are unaware of any branching process, are like the criticism of the Copernican theory that the mobility of the earth as a real physical fact is incompatible with the common sense interpretation of nature because we feel no such motion. In both cases the argument fails when it is shown that the theory itself predicts that our experience will be what it in fact is. (In the Copernican case the addition of Newtonian physics was required to be able to show that the earth's inhabitants would be unaware of any motion of the earth.)

have a requirement analogous to the "conservation of probability" in the classical case, we demand that the measure assigned to a trajectory at one time shall equal the sum of the measures of its separate branches at a later time. This is precisely the additivity requirement which we imposed and which leads uniquely to the choice of square-amplitude measure. Our procedure is therefore quite as justified as that of classical statistical mechanics.

Having deduced that there is a unique measure which will satisfy our requirements, the square-amplitude measure, we continue our deduction. This measure then assigns to the $i, j, \cdots k$th element of the superposition (24),

$$\phi_i^{S_1}\phi_j^{S_2}\cdots\phi_k^{S_r}\psi^{S_{r+1}}\cdots\psi^{S_n}\psi^{O}[\alpha_i^1,\alpha_j^2,\cdots\alpha_k^r] \tag{32}$$

the measure (weight)

$$M_{ij\ldots k}=(a_i a_j\cdots a_k)^*(a_i a_j\cdots a_k) \tag{33}$$

so that the observer state with memory configuration $[\alpha_i^1, \alpha_j^2, \cdots, \alpha_k^r]$ is assigned the measure $a_i^*a_i a_j^*a_j\cdots a_k^*a_k = M_{ij\ldots k}$. We see immediately that this is a product measure, namely,

$$M_{ij\ldots k}=M_i M_j\cdots M_k \tag{34}$$

where

$$M_l = a_l^* a_l$$

so that the measure assigned to a particular memory sequence $[\alpha_i^1, \alpha_j^2, \cdots, \alpha_k^r]$ is simply the product of the measures for the individual components of the memory sequence.

There is a direct correspondence of our measure structure to the probability theory of random sequences. *If we regard* the $M_{ij\ldots k}$ as probabilities for the sequences then the sequences are equivalent to the random sequences which are generated by ascribing to each term the *independent* probabilities $M_l = a_l^*a_l$. Now probability theory is equivalent to measure theory mathematically, so that we can make use of it, while keeping in mind that all results should be translated back to measure theoretic language.

Thus, in particular, if we consider the sequences to become longer and longer (more and more observations performed) *each* memory sequence of the final superposition will satisfy any given criterion for a randomly generated sequence, generated by the independent probabilities $a_l^*a_l$, except for a set of total measure which tends toward zero as the number of observations becomes unlimited. Hence all averages of functions over *any* memory sequence, including the special case of frequencies, can be computed from the probabilities $a_i^*a_i$, except for a set of memory sequences of measure zero. We have therefore shown that the statistical assertions of Process 1 will appear to be valid to the observer, *in almost all* elements of the superposition (24), in the limit as the number of observations goes to infinity.

While we have so far considered only sequences of observations of the same quantity upon identical systems, the result is equally true for arbitrary sequences of observations, as may be verified by writing more general sequences of measurements, and applying Rules 1 and 2 in the same manner as presented here.

We can therefore summarize the situation when the sequence of observations is arbitrary, when these observations are made upon the same or different systems in any order, and when the number of observations of each quantity in each system is very large, with the following result:

> Except for a set of memory sequences of measure nearly zero, the averages of any functions over a memory sequence can be calculated approximately by the use of the independent probabilities given by Process 1 for each initial observation, on a system, and by the use of the usual transition probabilities for succeeding observations upon the same system. In the limit, as the number of all types of observations goes to infinity the calculation is exact, and the exceptional set has measure zero.

This prescription for the calculation of averages over memory sequences by probabilities assigned to individual elements is precisely that of the conventional "external observation" theory (Process 1). Moreover, these predictions hold for almost all memory sequences. Therefore all predictions of the usual theory will appear to be valid to the observer in amost all observer states.

In particular, the uncertainty principle is never violated since the latest measurement upon a system supplies all possible information about the relative system state, so that there is no direct correlation between any earlier results of observation on the system, and the succeeding observation. Any observation of a quantity $B$, between two successive observations of quantity $A$ (all on the same system) will destroy the one-one correspondence between the earlier and later memory states for the result of $A$. Thus for alternating observations of different quantities there are fundamental limitations upon the correlations between memory states for the same observed quantity, these limitations expressing the content of the uncertainty principle.

As a final step one may investigate the consequences of allowing several observer systems to interact with (observe) the same object system, as well as to interact with one another (communicate). The latter interaction can be treated simply as an interaction which correlates parts of the memory configuration of one observer with another. When these observer systems are investigated, in the same manner as we have already presented in this section using Rules 1 and 2, one finds that in *all elements* of the final superposition:

1. When several observers have separately observed the same quantity in the object system and then communicated the results to one another they find that they

are in agreement. This agreement persists even when an observer performs his observation *after* the result has been communicated to him by another observer who has performed the observation.

2. Let one observer perform an observation of a quantity $A$ in the object system, then let a second perform an observation of a quantity $B$ in this object system which does not commute with $A$, and finally let the first observer repeat his observation of $A$. Then the memory system of the first observer will *not* in general show the same result for both observations. The intervening observation by the other observer of the noncommuting quantity $B$ prevents the possibility of any one to one correlation between the two observations of $A$.

3. Consider the case where the states of two object systems are correlated, but where the two systems do not interact. Let one observer perform a specified observation on the first system, then let another observer perform an observation on the second system, and finally let the first observer repeat his observation. Then it is found that the first observer always gets the same result both times, and the observation by the second observer has no effect whatsoever on the outcome of the first's observations. Fictitious paradoxes like that of Einstein, Podolsky, and Rosen[8] which are concerned with such correlated, noninteracting systems are easily investigated and clarified in the present scheme.

Many further combinations of several observers and systems can be studied within the present framework. The results of the present "relative state" formalism agree with those of the conventional "external observation" formalism in all those cases where that familiar machinery is applicable.

In conclusion, the continuous evolution of the state function of a composite system with time gives a complete mathematical model for processes that involve an idealized observer. When interaction occurs, the result of the evolution in time is a superposition of states, each element of which assigns a different state to the memory of the observer. Judged by the state of the memory in almost all of the observer states, the probabilistic conclusion of the usual "external observation" formulation of quantum theory are valid. In other words, pure Process 2 wave mechanics, without any initial probability assertions, leads to all the probability concepts of the familiar formalism.

### 6. DISCUSSION

The theory based on pure wave mechanics is a conceptually simple, causal theory, which gives predictions in accord with experience. It constitutes a framework in which one can investigate in detail, mathematically, and in a logically consistent manner a number of sometimes puzzling subjects, such as the measuring process itself and the interrelationship of several observers. Objections have been raised in the past to the conventional or "external observation" formulation of quantum theory on the grounds that its probabilistic features are postulated in advance instead of being derived from the theory itelf. We believe that the present "relative-state" formulation meets this objection, while retaining all of the content of the standard formulation.

While our theory ultimately justifies the use of the probabilistic interpretation as an aid to making practical predictions, it forms a broader frame in which to understand the consistency of that interpretation. In this respect it can be said to form a *metatheory* for the standard theory. It transcends the usual "external observation" formulation, however, in its ability to deal logically with questions of imperfect observation and approximate measurement.

The "relative state" formulation will apply to all forms of quantum mechanics which maintain the superposition principle. It may therefore prove a fruitful framework for the quantization of general relativity. The formalism invites one to construct the formal theory first, and to supply the statistical interpretation later. This method should be particularly useful for interpreting quantized unified field theories where there is no question of ever isolating observers and object systems. They all are represented in a *single* structure, the field. Any interpretative rules can probably only be deduced in and through the theory itself.

Aside from any possible practical advantages of the theory, it remains a matter of intellectual interest that the statistical assertions of the usual interpretation do not have the status of independent hypotheses, but are deducible (in the present sense) from the pure wave mechanics that starts completely free of statistical postulates.

---

[8] Einstein, Podolsky, and Rosen, Phys. Rev. 47, 777 (1935). For a thorough discussion of the physics of observation, see the chapter by N. Bohr in *Albert Einstein, Philosopher-Scientist* (The Library of Living Philosophers, Inc., Evanston, 1949).

Reprinted from REVIEWS OF MODERN PHYSICS, Vol. 29, No. 3, 463–465, July, 1957
Printed in U. S. A.

# Assessment of Everett's "Relative State" Formulation of Quantum Theory

JOHN A. WHEELER
*Palmer Physical Laboratory, Princeton University, Princeton, New Jersey*

THE preceding paper puts the principles of quantum mechanics in a new form.[1] Observations are treated as a special case of normal interactions that occur within a system, not as a new and different kind of process that takes place from without. The conventional mathematical formulation with its well-known postulates about probabilities of observations is derived as a *consequence* of the new or "*meta*" quantum mechanics. Both formulations apply as well to complex systems as to simple ones, and as well to particles as to fields. Both supply mathematical models for the physical world. In the new or "relative state" formalism this model associates with an isolated system a state function that obeys a linear wave equation. The theory deals with the totality of all the possible ways in which this state function can be decomposed into the sum of products of state functions for subsystems of the overall system—and nothing more. For example, in a system endowed with four degrees of freedom $x_1,x_2,x_3,x_4$, and a time coordinate, $t$, the general state can be written $\psi(x_1,x_2,x_3,x_4,t)$. However, there is *no* way in which $\psi$ defines any unique state for any subsystem (subset of $x_1,x_2,x_3,x_4$). The subsystem consisting of $x_1$ and $x_3$, say, *cannot* be assigned a state $u(x_1,x_3,t)$ independent of the state assigned to the subsystem $x_2$ and $x_4$. In other words, there is ordinarily *no* choice of $f$ or $u$ which will allow $\psi$ to be written in the form $\psi=u(x_1,x_3,t)f(x_2,x_4,t)$. The most that can be done is to associate a *relative* state to the subsystem, $u_{rel}(x_1,x_3,t)$, relative to some *specified* state $f(x_2,x_4,t)$ for the remainder of the system. The method of assigning relative states $u_{rel}(x_1,x_3,t)$ in one subsystem to specific states $f(x_2,x_4,t)$ for the remainder, permits one to decompose $\psi$ into a superposition of products, each consisting of one member of an orthonormal set for one subsystem and its corresponding relative state in the other subsystem:

$$\psi=\sum_i a_i f_i(x_2,x_4,t)u_{rel\,f_i}(x_1,x_3,t), \tag{1}$$

where $\{f\}$ is an orthonormal set. According as the functions $f_n$ constitute one or another family of orthonormal functions, the relative state functions $u_{rel\,f_n}$ have one or another dependence upon the variables of the remaining subsystem.

Another way of phrasing this unique association of relative state in one subsystem to states in the remainder is to say that the states are correlated. The totality of these correlations which can arise from all possible decompositions into states and relative states is all that can be read out of the mathematical model.

The model has a place for observations only insofar as they take place within the isolated system. The theory of observation becomes a special case of the theory of correlations between subsystems.

How does this mathematical model for nature relate to the present conceptual scheme of physics? Our conclusions can be stated very briefly: (1) The conceptual scheme of "relative state" quantum mechanics is completely different from the conceptual scheme of the conventional "external observation" form of quantum mechanics and (2) The conclusions from the new treatment correspond completely in familiar cases to the conclusions from the usual analysis. The rest of this note seeks to stress this *correspondence in conclusions* but also this *complete difference in concept*.

The "external observation" formulation of quantum mechanics has the great merit that it is dualistic. It associates a state function with the system under study —as for example a particle—but not with the *ultimate* observing equipment. The system under study can be enlarged to include the original object as a subsystem and also a piece of observing equipment—such as a Geiger counter—as another subsystem. At the same time the number of variables in the state function has to be enlarged accordingly. However, the *ultimate* observing equipment still lies outside the system that is treated by a wave equation. As Bohr[2] so clearly emphasizes, we always interpret the wave amplitude by way of observations of a classical character made from outside the quantum system. The conventional formalism admits no other way of interpreting the wave amplitude; it is logically self-consistent; and it rightly rules out any classical description of the internal dynamics of the system. With the help of the principle of complementarity the "external observation" formulation nevertheless keeps all it consistently can of classical concepts. Without this possibility of classical measuring equipment the mathematical machinery of quantum mechanics would seem at first sight to admit no correlation with the physical world.

Instead of founding quantum mechanics upon classical physics, the "relative state" formulation uses a completely different kind of model for physics. This new model has a character all of its own; is conceptually

[1] Hugh Everett, III, Revs. Modern Phys. 29, 454 (1957).

[2] Chapter by Niels Bohr in *Albert Einstein, Philosopher-Scientist*, edited by P. A. Schilpp (The Library of Living Philosophers, Inc., Evanston, Illinois, 1949).

self-contained; defines its own possibilities for interpretation; and does not require for its formulation any reference to classical concepts. It is difficult to make clear how decisively the "relative state" formulation drops classical concepts. One's initial unhappiness at this step can be matched but few times in history[3]: when Newton described gravity by anything so preposterous as action at a distance; when Maxwell described anything as natural as action at a distance in terms as unnatural as field theory; when Einstein denied a privileged character to any coordinate system, and the whole foundations of physical measurement at first sight seemed to collapse. How can one consider seriously a model for nature that follows neither the Newtonian scheme, in which coordinates are functions of time, nor the "external observation" description, where probabilities are ascribed to the possible outcomes of a measurement? Merely to analyze the alternative decompositions of a state function, as in (1), without saying what the decomposition means or how to interpret it, is apparently to define a theoretical structure almost as poorly as possible! Nothing quite comparable can be cited from the rest of physics except the principle in general relativity that all regular coordinate systems are equally justified. As in general relativity, so in the relative-state formulation of quantum mechanics the analysis of observation is the key to the physical interpretation.

Observations are not made from outside the system by some super-observer. There is no observer on hand to use the conventional "external observation" theory. Instead, the whole of the observer apparatus is treated in the mathematical model as part of an isolated system. All that the model will say or ever can say about observers is contained in the interrelations of eigenfunctions for the object part of this isolated system and relative state functions of the remaining part of the system. Every attempt to ascribe probabilities to observables is as out of place in the relative state formalism as it would be in any kind of quantum physics to ascribe coordinate and momentum to a particle at the same time. The word "probability" implies the notion of observation from outside with equipment that will be described typically in classical terms. Neither these classical terms, nor observation from outside, nor a priori probability considerations come into the *foundations* of the relative state form of quantum theory.

So much for the conceptual differences between the new and old formulations. Now for their correspondence. The preceding paper shows that this correspondence is detailed and close. The tracing out of the correspondence demands that the system include something that can be called an observing subsystem. This subsystem can be as simple as a particle which is to collide with a particle that is under study. In this case the correspondence occurs at a primitive level between the relative state formalism where the system consists of two particles, and the external observation theory where the system consists of only one particle. The correlations between the eigenfunctions of the object particle and the relative state functions of the observer particle in the one scheme are closely related in the other scheme to the familiar statements about the relative probabilities for various possible outcomes of a measurement on the object particle.

A more detailed correspondence can be traced between the two forms of quantum theory when the observing system is sufficiently complex to have what can be described as memory states. In this case one can see the complementary aspects of the usual external observation theory coming into evidence in another way in the relative state theory. They are expressed in terms of limitations on the degree of correlation between the memory states for successive observations on a system of the same quantity, when there has been an intervening observation of a noncommuting quantity. In this sense one has in the relative state formalism for the first time the possibility of a closed mathematical model for complementarity.

In physics it is not enough for a single observer or apparatus to make measurements. Different pieces of equipment that make the same type of measurement on the same object system must show a pattern of consistency if the concept of measurement is to make sense. Does not such consistency demand the external observation formulation of quantum theory? There the results of the measurements can be spelled out in classical language. Is not such "language" a prerequisite for comparing the measurements made by different observing systems?

The analysis of multiple observers in the preceding paper by the theory of relative states indicates that the necessary consistency between measurements is already obtained without going to the external observer formulation. To describe this situation one can use if he will the words "communication in clear terms always demands classical concepts." However, the kind of physics that goes on does not adjust itself to the available terminology; the terminology has to adjust itself in accordance with the kind of physics that goes on. In brief, the problem of multiple observers solves itself within the theory of relative states, not by adding the conventional theory of measurement to that theory.

It would be too much to hope that this brief survey should put the relative state formulation of quantum theory into completely clear focus. One can at any rate end by saying what it does not do. It does not seek to supplant the conventional external observer formalism, but to give a new and independent foundation for that formalism. It does not introduce the idea of a super-observer; it rejects that concept from the start. It does not supply a prescription to say what is the correct

[3] See, for example, Philipp Frank's *Modern Science and Its Philosophy* (George Braziller, New York, 1955), Chap. 12.

functional form of the Hamiltonian of any given system. Neither does it supply any prediction as to the functional dependence of the over-all state function of the isolated system upon the variables of the system. But neither does the classical universe of Laplace supply any prescription for the original positions and velocities of all the particles whose future behavior Laplace stood ready to predict. In other words, the relative state theory does not pretend to answer all the questions of physics. The concept of relative state does demand a totally new view of the foundational character of physics. No escape seems possible from this relative state formulation if one wants to have a complete mathematical model for the quantum mechanics that is internal to an isolated system. Apart from Everett's concept of relative states, no self-consistent system of ideas is at hand to explain what one shall mean by quantizing[4] a closed system like the universe of general relativity.

[4] C. W. Misner, Revs. Modern Phys. 29, 497 (1957).

Reprinted from Physics Today, Vol. 23, No. 9 (September 1970).

# Quantum mechanics and reality

Could the solution to the dilemma of indeterminism be a universe in which all possible outcomes of an experiment actually occur?

Bryce S. DeWitt

Despite its enormous practical success, quantum theory is so contrary to intuition that, even after 45 years, the experts themselves still do not all agree what to make of it. The area of disagreement centers primarily around the problem of describing observations. Formally, the result of a measurement is a superposition of vectors, each representing the quantity being observed as having one of its possible values. The question that has to be answered is how this superposition can be reconciled with the fact that in practice we only observe one value. How is the measuring instrument prodded into making up its mind which value it has observed?

Of the three main proposals for solving this dilemma, I shall focus on one that pictures the universe as continually splitting into a multiplicity of mutually unobservable but equally real worlds, in each one of which a measurement does give a definite result. Although this proposal leads to a bizarre world view, it may be the most satisfying answer yet advanced.

### Quantum theory of measurement

In its simplest form the quantum theory of measurement considers a world composed of just two dynamical entities, a *system* and an *apparatus*. Both are subject to quantum-mechanical laws, and hence one may form a combined state vector that can be expanded in terms of an orthonormal set of basis vectors

$$|s,A\rangle = |s\rangle|A\rangle \qquad (1)$$

where $s$ is an eigenvalue of some system observable and $A$ is an eigenvalue of some apparatus observable. (Additional labels have been suppressed for simplicity.) The Cartesian product structure of equation 1 reflects an implicit assumption that, under appropriate conditions, such as the absence of coupling, the system and apparatus can act as if they are isolated, independent and distinguishable. It is also convenient to assume that the eigenvalue $s$ ranges over a discrete set while the eigenvalue $A$ ranges over a continuum.

Suppose that the state of the world at some initial instant is represented by a normalized vector of the form

$$|\Psi_0\rangle = |\psi\rangle|\Phi\rangle \qquad (2)$$

where $|\psi\rangle$ refers to the system and $|\Phi\rangle$ to the apparatus. In such a state the system and apparatus are said to be "uncorrelated." For the apparatus to learn something about the system the two must be coupled together for a certain period, so that their combined state will not retain the form of equation 2 as time passes. The final result of the coupling will be described by the action of a certain unitary operator U

Bryce DeWitt is professor of physics at the University of North Carolina.

**Schrödinger's cat.** The animal trapped in a room together with a Geiger counter and a hammer, which, upon discharge of the counter, smashes a flask af prussic acid. The counter contains a trace of radioactive material—just enough that in one hour there is a 50% chance one of the nuclei will decay and therefore an equal chance the cat will be poisoned. At the end of the hour the total wave function for the system will have a form in which the living cat and the dead cat are mixed in equal portions. Schrödinger felt that the wave mechanics that led to this paradox presented an unacceptable description of reality. However, Everett, Wheeler and Graham's interpretation of quantum mechanics pictures the cats as inhabiting two simultaneous, noninteracting, but equally real worlds.

$$|\Psi_1\rangle = U|\Psi_0\rangle \quad (3)$$

Because the apparatus observes the system and not vice versa, we must choose a coupling operator U that reflects this separation of function. Let U have the following action on the basis vectors defined in equation 1 (or on some similar basis):

$$U|s,A\rangle = |s,A + gs\rangle = |s\rangle|A + gs\rangle \quad (4)$$

Here, $g$ is a coupling constant, which may be assumed to be adjustable. If the initial state of the system were $|s\rangle$ and that of the apparatus were $|A\rangle$ then this coupling would be said to result in an "observation," by the apparatus, that the system observable has the value $s$. This observation or "measurement," would be regarded as "stored" in the apparatus "memory" by virtue of the permanent shift from $|A\rangle$ to $|A + gs\rangle$ in the apparatus state vector.

## Is this definition adequate?

This particular choice for U, essentially formulated by John von Neumann,[1] is frequently criticized because it is not sufficiently general and because it artificially delimits the concept of measurement. Some writers[2] have also insisted that the process described by equation 4 merely prepares the system and that the measurement is not complete until a more complicated piece of apparatus observes the outcome of the preparation.

It is perfectly true that laboratory measurements are much more complicated than that described by equation 4 and often involve interactions that do not establish precise correlations between pairs of observables such as $s$ and $A$. However, apart from such noncorrelative interactions, every laboratory measurement consists of one or more sequences of interactions, each essentially of the von Neumann type. Although it is only the results of the final interactions with the recording devices that we usually regard as being stored, each von Neumann-type "apparatus" in every sequence leading to a final interaction may itself be said to possess a memory, at least momentarily. This memory differs in no fundamental way from that of the sophisticated automaton (apparatus-plus-memory sequence) at the end of the line. It is the elementary component that must be understood if we are to understand quantum mechanics itself.

In his original analysis of the measurement process,[1] von Neumann assumed that the coupling between system and apparatus leaves the system observable $s$ undisturbed. Most of his conclusions would have remained unaffected had he removed this restriction, and we are not making such an assumption here. Although measurements of the nondisturbing type do exist, more frequently the observable suffers a change. It can nevertheless be shown[3] that if suitable devices are used, such as the compensation devices introduced by Niels Bohr and Leon Rosenfeld in their analysis of electromagnetic-field measurements,[4] the apparatus can record what the value of the system observable would have been without the coupling. For this reason, we work in a modified version of the so-called "interaction picture," in which only that part of the state vector that refers to the apparatus changes during the coupling interval.

If the coupling is known, the hypothetical undisturbed system observable may be expressed in terms of the actual dynamical variables of system *plus* apparatus. Hence, the operator of which this observable is an eigenvalue is not itself hypothetical, and no inconsistency **will arise if we take it to be the observable to which the label *s* refers on the right side of equation 4.**

## Infinite regression

Consider now what happens to the initial state vector in equation 2 as a result of the measurement process of equation 4. Using the orthonormality and assumed completeness of the basis vectors, we easily find that

$$|\Psi_1\rangle = \sum_s c_s|s\rangle|\Phi[s]\rangle \qquad (5)$$

where $= \sum_s \langle s|\psi\rangle |s\rangle |\Phi[s]\rangle$

$$c_s = \langle s|\psi\rangle \qquad (6)$$

$$|\Phi[s]\rangle = \int |A + gs\rangle\Phi(A)dA \qquad (7)$$

$$\Phi(A) = \langle A|\Phi\rangle \qquad (8)$$

The final state vector in equation 5 does not represent the system observable as having any unique value—unless, of course, $|\psi\rangle$ happens to be one of the basis vectors $|s\rangle$. Rather it is a linear superposition of vectors $|s\rangle|\Phi[s]\rangle$, each of which represents the system observable as having assumed one of its possible values and the apparatus as having observed that value. For each possibility the observation will be a good one, that is, capable of distinguishing adjacent values of $s$, provided

$$\Delta A \ll g\Delta s \qquad (9)$$

where $\Delta s$ is the spacing between adjacent values and $\Delta A$ is the variance in $A$ about its mean value relative to the distribution function $|\Phi(A)|^2$. Under these conditions we have

$$\langle\Phi[s]|\Phi[s']\rangle = \delta_{ss'} \qquad (10)$$

In other words, the wave function of the apparatus takes the form of a packet that is initially single but subsequently splits, as a result of the coupling to the system, into a multitude of mutually orthogonal packets, one for each value of $s$.

Here the controversies over the interpretation of quantum mechanics start. For most people, a state like that of equation 5 does not represent the actual occurrence of an observation. They conceive the apparatus to have entered a kind of schizophrenic state in which it is unable to decide what value it has found for the system observable. At the same time they can not deny that the coupling chosen between system and apparatus would, in the classical theory, have led to a definite outcome. They therefore face a crisis. How can they prod the apparatus into making up its mind?

The usual suggestion is to introduce a second apparatus to get at the facts simply by looking at the first apparatus to see what it has recorded. But an analysis carried out along the above lines quickly shows that the second apparatus performs no better than the first. It too goes into a state of schizophrenia. The same thing happens with a third apparatus, and a fourth, and so on. This chain, known as "von Neumann's catastrophe of infinite regression," only makes the crisis worse.

**Change the rules**

There are essentially three distinct ways of getting out of the crisis. The first is to change the rules of the game by changing the theory, the object being to break von Neumann's infinite chain. Eugene Wigner is the most distinguished proponent of this method. Taking a remarkably anthropocentric stand, he proposes that the entry of the measurement signal into the consciousness of an observer is what triggers the decision and breaks the chain.[5] Certainly the chain had better be broken at this point, as the human brain is usually where laboratory-measurement sequences terminate. One is reminded of the sign that used to stand on President Truman's desk: "The buck stops here."

Wigner does not indulge in mere handwaving; he actually sketches a possible mathematical description of the conversion from a pure to a mixed state, which might come about as a result of the grossly nonlinear departures from the normal Schrödinger equation that he believes must occur when conscious beings enter the picture. He also proposes that a search be made for unusual effects of consciousness acting on matter.[5]

Another proponent of the change-the-rules method is David Bohm.[6,7] Unlike Wigner, who does not wish to change the theory below the level of consciousness, Bohm and his school want to change the foundations so that even the first apparatus is cured of its schizophrenia. This they do by introducing so-called "hidden variables." Whatever else may be said of hidden-variable theories, it must be admitted that they do what they are supposed to. The first such theory[6] in fact worked too well; there was no way of distinguishing it experimentally from conventional quantum mechanics. More recent hid-

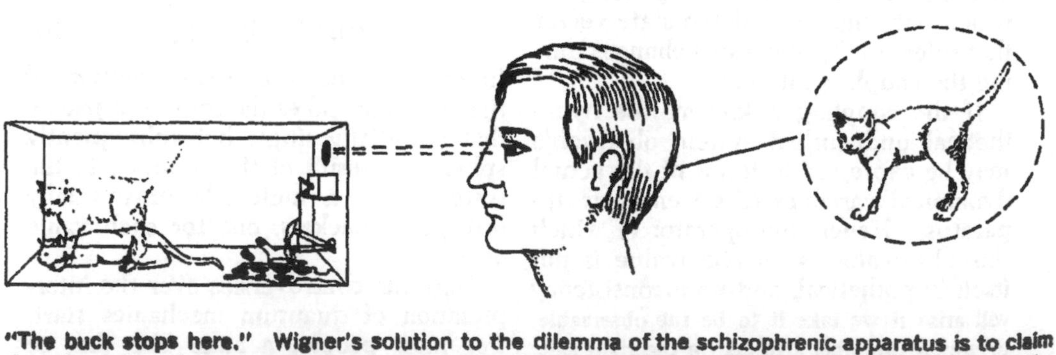

**"The buck stops here."** **Wigner's solution to the dilemma of the schizophrenic apparatus is to claim that the entry of the measurement signal into the consciousness of a human observer triggers the decision as to which of the possible outcomes is observed—that is, whether the cat is alive or dead.**

den-variable theories are susceptible to possible experimental verification (or disproof).[7]

## The Copenhagen collapse

The second method of escaping the von Neumann catastrophe is to accept the so-called "conventional," or "Copenhagen," interpretation of quantum mechanics. (Reference 8 contains a selected list of papers on this topic.) In speaking of the adherents of this interpretation it is important to distinguish the active adherents from the rest, and to realize that even most textbook authors are not included among the former. If a poll were conducted among physicists, the majority would profess membership in the conventionalist camp, just as most Americans would claim to believe in the Bill of Rights, whether they had ever read it or not. The great difficulty in dealing with the activists in this camp is that they too change the rules of the game but, unlike Wigner and Bohm, pretend that they don't.

According to the Copenhagen interpretation of quantum mechanics, whenever a state vector attains a form like that in equation 5 it immediately collapses. The wave function, instead of consisting of a multitude of packets, reduces to a single packet, and the vector $|\Psi_1\rangle$ reduces to a corresponding element $|s\rangle|\Phi[s]$ of the superposition. To which element of the superposition it reduces one can not say. One instead assigns a probability distribution to the possible outcomes, with weights given by

$$w_s = |c_s|^2 \qquad (11)$$

The collapse of the state vector and the assignment of statistical weights do not follow from the Schrödinger equation, which generates the operator U (equation 4). They are consequences of an external *a priori* metaphysics, which is allowed to intervene at this point and suspend the Schrödinger equation, or rather replace the boundary conditions on its solution by those of the collapsed state vector. Bohm and Wigner try to construct explicit mechanisms for bringing about the collapse, but the conventionalists claim that it does not matter how the state vector is collapsed. To them the state vector does not represent reality but only an algorithm for making statistical predictions. In fact, if the measurement involves a von Neumann chain they are even willing to leave the state vector uncollapsed over an arbitrary number of links, just so long as it is treated as collapsed somewhere along the line.

The Copenhagen view promotes the impression that the collapse of the state vector, and even the state vector itself, is all in the mind. If this impression is correct, then what becomes of reality? How can one treat so cavalierly the objective world that obviously exists all around us? Einstein, who opposed to his death the metaphysical solution of the Copenhagen school, must surely have expressed himself thus in his moments of private indignation over the quantum theory. I am convinced that these sentiments also underlie much of the current dissatisfaction with the conventional interpretation of quantum mechanics.

## Historical interpretations

This problem of the physical interpretation of the quantum theory haunted its earliest designers. In 1925 and 1926 Werner Heisenberg had just succeeded in breaking the quantum theory from its moorings to the old quantum rules. Through the work of Max Born, Pascual Jordan, Erwin Schrödinger, P. A. M. Dirac and Heisenberg himself, this theory soon acquired a fully developed mathematical formalism. The challenge then arose of elucidating the physical interpretation of this formalism independently of anything that had gone on before.

Heisenberg attempted to meet this challenge by inventing numerous thought experiments, each of which was subjected to the question: "Can it be described by the formalism?" He conjectured that the set of experiments for which the answer is "yes" is identical to the set permitted by nature.[9] To put the question in its most extreme form in each case meant describing the complete experiment, including the measuring apparatus itself, in quantum-mechanical terms.

At this point Bohr entered the picture and deflected Heisenberg somewhat from his original program. Bohr convinced Heisenberg and most other physicists that quantum mechanics has no meaning in the absence of a classical realm capable of unambiguously recording the results of observations. The mixture of metaphysics with physics, which this notion entailed, led to the almost universal belief that the chief issues of interpretation are epistemological rather than ontological: The quantum realm must be viewed as a kind of ghostly world whose symbols, such as the wave function, represent potentiality rather than reality.

## The EWG metatheorem

What if we forgot all metaphysical ideas and started over again at the point where Heisenberg found himself in 1925? Of course we can not forget everything; we will inevitably use 45 years of hindsight in attempting to restructure our interpretation of quantum mechanics. Let us nevertheless try

- to take the mathematical formalism of quantum mechanics as it stands without adding anything to it
- to deny the existence of a separate classical realm
- to assert that the state vector never collapses.

In other words, what if we assert that the formalism is all, that nothing else is needed? Can we get away with it? The answer is that we can. The proof of this assertion was first given in 1957 by Hugh Everett[10] with the encouragement of John Wheeler[11] and has been subsequently elaborated by R. Neill Graham.[12] It constitutes the third way of getting out of the crisis posed by the catastrophe of infinite regression.

Everett, Wheeler and Graham (EWG) postulate that the real world, or any isolated part of it one may wish for the moment to regard as *the* world, is faithfully represented solely by the *following mathematical objects:* a vector in a Hilbert space; a set of dynamical equations (derived from a variational principle) for a set of operators that act on the Hilbert space, and a set of commutation relations for the operators (derived from the Poisson brackets of the classical theory by the quantization rule, where classical analogs exist). Only one additional postulate is then needed to give physical meaning to the mathematics. This is the postulate of complexity: The world must be sufficiently complicated that it be decomposable into systems and apparatuses.

Without drawing on any external metaphysics or mathematics other than the standard rules of logic, EWG are able, from these postulates, to prove the following metatheorem: *The mathematical formalism of the quantum theory is capable of yielding its own interpretation.* To prove this metatheorem, EWG must answer two questions:

- How can the conventional probability interpretation of quantum mechanics emerge from the formalism itself?
- How can any correspondence with reality be achieved if the state vector never collapses?

## Absolute chance

Before giving the answers to these questions, let us note that the conventional interpretation of quantum mechanics confuses two concepts that really ought to be kept distinct—probability as it relates to quantum mechanics and probability as it is understood in statistical mechanics. Quantum mechanics is a theory that attempts to describe in mathematical language a world in which chance is not a measure of our ignorance but is absolute. It must inevitably lead to states, like that of equation 5, that undergo multiple fission, corresponding to the many possible outcomes of a given measurement. Such behavior is built into the formalism. However, precisely because quantum-mechanical chance is *not* a measure of our ignorance, we ought not to tamper with the state vector merely because we acquire new information as a result of a measurement.

The obstacle to taking such a lofty view of things, of course, is that it forces us to believe in the reality of all the simultaneous worlds represented in the superposition described by equation 5, in each of which the measurement has yielded a different outcome. Nevertheless, this is precisely what EWG would have us believe. According to them the real universe is faithfully represented by a state vector similar to that in equation 5 but of vastly greater complexity. This universe is constantly splitting into a stupendous number of branches, all resulting from the measurementlike interactions between its myriads of components. Moreover, every quantum transition taking place on every star, in every galaxy, in every remote corner of the universe is splitting our local world on earth into myriads of copies of itself.

## A splitting universe

I still recall vividly the shock I experienced on first encountering this multiworld concept. The idea of $10^{100}+$ slightly imperfect copies of oneself all constantly splitting into further copies, which ultimately become unrecognizable, is not easy to reconcile with common sense. Here is schizophrenia with a vengeance. How pale in comparison is the mental state of the imaginary friend, described by Wigner,[5] who is hanging in suspended animation between only *two* possible outcomes of a quantum measurement. Here we must surely protest. None of us feels like Wigner's friend. We do not split in two, let alone into $10^{100}+$! To this EWG reply: To the extent that we can be regarded simply as automata and hence on a par with ordinary measuring apparatuses, the laws of quantum mechanics do not allow us to feel the splits.

A good way to prove this assertion is to begin by asking what would happen, in the case of the measurement described earlier by equations 4 and 5, if one introduced a second apparatus that not only looks at the memory bank of the first apparatus but also carries out an independent direct check on the value of the system observable. If the splitting of the universe is to be unobservable the results had better agree.

The couplings necessary to accomplish the desired measurements are readily set up. The final result is as follows (see reference 13): The state vector at the end of the coupling interval again takes the form of a linear superposition of vectors, each of which represents the system observable as having assumed one of its possible values. Although the value varies from one element of the superposition to another, not only do both apparatuses within a given element observe the value appropriate to that element, but also, by straightforward communication, they agree that the results of their observations are identical. The splitting into branches is thus unobserved.

## Probability interpretation

We must still discuss the questions of the coefficients $c_n$ in equations 5 and 6. EWG give no *a priori* interpretation to these coefficients. In order to find an interpretation they introduce an apparatus that makes repeated measure-

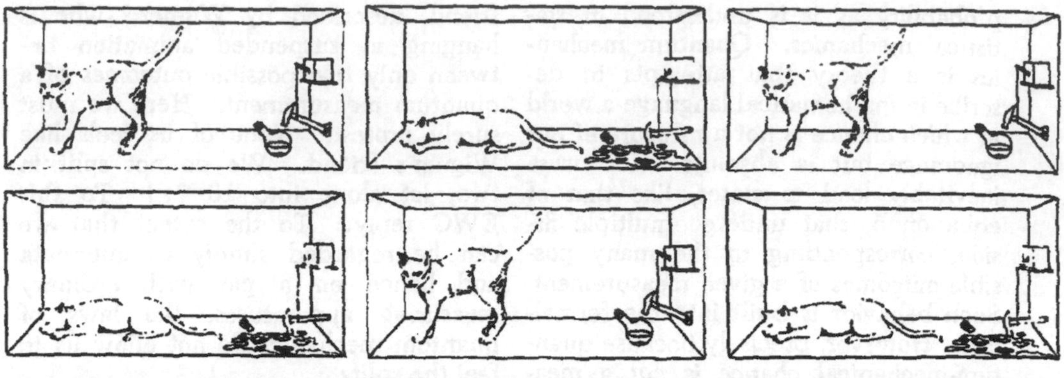

**The Copenhagen collapse. This interpretation pictures the total wave function as collapsing to one state of the superposition and assigns a probability that the wave function will collapse to a given state. *Only for repetition on an ensemble of cats would live and dead cats be equally real.***

ments on an ensemble of identical systems in identical states. The initial state then has the form

$$|\Psi_0\rangle = |\psi_1\rangle|\psi_2\rangle\ldots|\Phi\rangle \quad (12)$$

where

$$\langle s|\psi_i\rangle = c_s \quad \text{for all } i \quad (13)$$

and the successive measurements are described in terms of basis vectors

$$|s_1\rangle|s_2\rangle\ldots|A_1,A_2,\ldots\rangle \quad (14)$$

If the apparatus observes each system exactly once, in sequence, then the $n$th measurement is represented by a unitary transition of the form

$$U_n(|s_1\rangle|s_2\rangle\ldots|A_1,A_2,\ldots,A_n,\ldots\rangle = |s_1\rangle|s_2\rangle\ldots|A_1,A_2,\ldots,A_n+gs_n,\ldots\rangle \quad (15)$$

After $N$ measurements the state vector in equation 12 is changed to

$$|\Psi_N\rangle = \sum_{s_1,s_2\ldots} c_{s_1}c_{s_2}\ldots|s_1\rangle|s_2\rangle\ldots|\Phi[s_1,s_2\ldots s_N]\rangle \quad (16)$$

where

$$|\Phi[s_1,s_2\ldots]\rangle = \int dA_1\int dA_2\ldots|A_1+gs_1,A_2+gs_2,\ldots\rangle\,\Phi(A_1,A_2\ldots) \quad (17)$$

$$\Phi(A_1,A_2\ldots) = \langle A_1,A_2\ldots|\Phi\rangle \quad (18)$$

Although every system is initially in exactly the same state as every other, the apparatus does not generally record a sequence of identical values for the system observable, even within a single element of the superposition of equation 16. Each memory sequence $s_1,s_2,\ldots s_N$ yields a certain distribution of possible values for the system observable, and each distribution may be subjected to a statistical analysis. The first and simplest part of such an analysis is the calculation of the relative frequency function of the distribution:

$$f(s;s_1\ldots s_N) = \frac{1}{N}\sum_{n=1}^{N}\delta_{ss_n} \quad (19)$$

Let us introduce the function

$$\delta(s_1\ldots s_N) = \sum_s [f(s;s_1\ldots s_N) - w_s]^2 \quad (20)$$

where the $w$'s are any positive numbers that add up to unity. This is the first of a hierarchy of functions that measure the degree to which the sequence $s_1\ldots s_N$ deviates from a random sequence with weights $w_s$. Let us choose for the $w$'s the numbers defined in equation 11, and let us introduce an arbitrarily small positive number $\epsilon$. We shall call the sequence $s_1\ldots s_N$ "first random" if $\delta(s_1\ldots s_N) < \epsilon$ and "non-first-random" otherwise.

Suppose now we remove from the superposition of equation 16 all those elements for which the apparatus memory sequence is non-first-random. Denote the result by $|\Psi_N^\epsilon\rangle$. This vector has the remarkable property that it differs negligibly from $|\Psi_N\rangle$ in the limit $N \to \infty$. More precisely,

$$\lim_{N\to\infty} (|\Psi_N\rangle - |\Psi_N^\epsilon\rangle) = 0 \quad \text{for all } \epsilon > 0 \quad (21)$$

A proof will be found in reference 13.

A similar result is obtained if $|\Psi_N^\epsilon\rangle$ is redefined by excluding, in addition, elements of the superposition whose memory sequences fail to meet any finite combination of the infinity of other requirements for a random sequence. The conventional probability interpretation of quantum mechanics thus emerges from the formalism itself. Nonrandom memory sequences in equation 16 are of measure zero in the Hilbert space, in the limit as $N$ goes to infinity. Each automaton in the superposition sees the world obeying the familiar statistical quantum laws. However, there exists no outside agency that can designate which branch of the superposition is to be regarded as the real world. All are equally real, and yet each is unaware of the others. These conclusions obviously admit of immediate extension to the world of cosmology. Its state vector is like a tree with an enormous number of branches. Each branch corresponds to a possible universe-as-we-actually-see-it.

## Maverick worlds

The alert reader may now object that the above argument is circular, that in order to derive the *physical* probability interpretation of quantum mechanics, based on sequences of observations, we have introduced a *nonphysical* probability concept, namely that of the measure of a subspace in Hilbert space. This concept is alien to experimental physics because it involves many elements of the superposition at once, and hence many simultaneous worlds, that are supposed to be unaware of one another.

The problem that this objection raises is like many that have arisen in the long history of probability theory. Actually, EWG do not in the end exclude any element of the superposition. All the worlds are there, even those in which everything goes wrong and all the statistical laws break down. The situation is no different from that which we face in ordinary statistical mechanics. If the initial conditions were right, the universe-as-we-see-it could be a place in which heat sometimes flows from cold bodies to hot. We can perhaps argue that in those branches in which the universe makes a habit of misbehaving in this way, life fails to evolve; so no intelligent automata are around to be amazed by it.

It is also possible that maverick worlds are simply absent from the grand superposition. This could be the case if ordinary three-space is compact and the universe is finite. The wave function of a finite universe must itself contain only a finite number of branches. It simply may not have enough fine structure to accommodate maverick worlds. The extreme smallness of the portion of Hilbert space that such worlds would have to occupy becomes obvious when one compares the length of a Poincaré cycle, for even a small portion of the universe, to a typical cosmological time scale.

## Questions of practicality

The concept of a universal wave function leads to important questions regarding the practical application of the quantum-mechanical formalism. If I am part of the universe, how does it happen that I am able, without running into inconsistencies, to include as much or as little as I like of the real world of cosmology in my state vector? Why should I be so fortunate as to be able, in practice, to avoid dealing with the state vector of the universe?

The answer to these questions is to be found in the statistical implications of sequences of measurements of the kind that led us to the state vector of equation 16. Consider one of the memory sequences in this state vector. This memory sequence defines an average value for the system observable, given by

$$\langle s\rangle_{s_1\ldots N} = \sum_s sf(s;s_1\ldots s_N) \qquad (22)$$

If the sequence is random, as it is increasingly likely to be when $N$ becomes large, this average will differ only by an amount of order $\epsilon$ from the average

$$\langle s\rangle = \sum_s sw_s \qquad (23)$$

But the latter average may also be expressed in the form

$$\langle s\rangle = \langle\psi|\mathbf{s}|\psi\rangle \qquad (24)$$

where $|\psi\rangle$ is the initial state vector of any one of the identical systems and $\mathbf{s}$ is the operator of which the $s$'s are the eigenvalues. In this form the basis vectors $|s\rangle$ do not appear. Had we chosen to introduce a different apparatus, designed to measure some observable $\mathbf{r}$ not equal to $\mathbf{s}$, a sequence of repeated measurements would have yielded in this case an average approximately equal to

$$\langle r\rangle = \langle\psi|\mathbf{r}|\psi\rangle \qquad (25)$$

In terms of the basis vectors $|s\rangle$ this average is given by

$$\langle r\rangle = \sum_{s,s'} c_s^*\langle s|\mathbf{r}|s'\rangle c_{s'} \qquad (26)$$

Now suppose that we first measure $\mathbf{s}$ and then perform a statistical analysis on $\mathbf{r}$. We introduce a second apparatus that performs a sequence of observations on a set of identical two-component systems all in identical states given by the vector $|\Psi_1\rangle$ of equation 5. Each of the latter systems is composed of one of the original systems together with an apparatus that has just measured the observable $\mathbf{s}$. In view of the packet orthogonality relations, given by equation 10, we shall find for the average of $\mathbf{r}$ in this case

$$\langle r\rangle = \langle\Psi_1|\mathbf{r}|\Psi_1\rangle = \sum_s w_s\langle s|\mathbf{r}|s\rangle \qquad (27)$$

The averages in equations 26 and 27 are generally not equal. In equation 27, the measurement of $s$, which the first apparatus has performed, has destroyed the quantum interference effects that are still present in equation 26. Thus the elements of the superposition in equation 5 may be treated as if they were members of a statistical ensemble.

This result is what allows us, in practice, to collapse the state vector after a measurement has occurred, and to use the techniques of ordinary statistical mechanics, in which we change the boundary conditions upon receipt of new information. It is also what permits us to introduce systems having well defined initial states, without at the same time introducing the apparatuses that prepared the systems in those states. In brief, it is what allows us to start at any point in any branch of the universal state vector without worrying about previous or simultaneous branches.

We may, in principle, restore the interference effects of equation 26 by bringing the apparatus packets back together again. But then the correlations between system and apparatus are destroyed, the apparatus memory is wiped out and no measurement results. If one attempts to maintain the correlations by sneaking in a second apparatus to "have a look" before the packets are brought back together, then the state vector of the second apparatus must be introduced, and the separation of *its* packets will destroy the interference effects.

### Final assessment

Clearly the EWG view of quantum mechanics leads to experimental predictions identical with those of the Copenhagen view. This, of course, is its major weakness. Like the original Bohm theory[6] it can never receive operational support in the laboratory. No experiment can reveal the existence of the "other worlds" in a superposition like that in equations 5 and 16. However, the EWG theory does have the pedagogical merit of bringing most of the fundamental issues of measurement theory clearly into the foreground, and hence of providing a useful framework for discussion.

Moreover a decision between the two interpretations may ultimately be made on grounds other than direct laboratory experimentation. For example, in the very early moments of the universe, during the cosmological "Big Bang," the universal wave function may have possessed an overall coherence as yet unimpaired by condensation into noninterfering branches. Such initial coherence may have testable implications for cosmology.

Finally, the EWG interpretation of quantum mechanics has an important contribution to make to the philosophy of science. By showing that formalism alone is sufficient to generate interpretation, it has breathed new life into the old idea of a direct correspondence between formalism and reality. The reality implied here is admittedly bizarre. To anyone who is awestruck by the vastness of the presently known universe, the view from where Everett, Wheeler and Graham sit is truly impressive. Yet it is a completely causal view, which even Einstein might have accepted. At any rate, it has a better claim than most to be the natural end product of the interpretation program begun by Heisenberg in 1925.

## References


1. J. von Neumann, *Mathematical Foundations of Quantum Mechanics*, Princeton University Press, Princeton (1955).
2. H. Margenau, Phil. Sci. 4, 337 (1937); PHYSICS TODAY 7, no. 10, 6 (1954).
3. B. S. DeWitt, *Dynamical Theory of Groups and Fields*, Gordon and Breach, New York (1965), pp. 16–29.
4. N. Bohr, L. Rosenfeld, Kgl. Danske Videnskab. Selskab, Mat.-Fys. Medd. 12, no. 8 (1933).
5. E. P. Wigner, "Remarks on the Mind-Body Question," in *The Scientist Speculates* (I. J. Good, ed), William Heinemann Ltd, London (1961). Reprinted in E. P. Wigner, *Symmetries and Reflections*, Indiana University Press, Bloomington (1967).
6. D. Bohm, Phys. Rev. 85, 166 and 180 (1952); 87, 389 (1952); 89, 319 and 458 (1953).
7. D. Bohm, J. Bub, Rev. Mod. Phys. 38, 453 and 470 (1966).
8. A. Petersen, *Quantum Physics and the Philosophical Tradition,* MIT Press, Cambridge (1968).
9. W. Heisenberg, "Quantum Theory and Its Interpretation," in *Niels Bohr* (S. Rozental, ed), North Holland, Wiley, New York (1967).
10. H. Everett III, Rev. Mod. Phys. 29, 454 (1957).
11. J. A. Wheeler, Rev. Mod. Phys. 29, 463 (1957).
12. R. N. Graham, PhD thesis, University of North Carolina (in preparation).
13. B. S. DeWitt, "The Everett-Wheeler Interpretation of Quantum Mechanics," in *Battelle Rencontres, 1967 Lectures in Mathematics and Physics* (C. DeWitt, J. A. Wheeler, eds), W. A. Benjamin Inc., New York (1968). □

Reprinted From
*Foundations of Quantum Mechanics*
© 1971, IL Corso
*Academic Press Inc. - New York*

# The Many-Universes Interpretation of Quantum Mechanics (*).

B. S. De Witt
*Department of Physics, University of North Carolina at Chapel Hill*

## Introduction.

Although forty five years have passed since Heisenberg first unlocked the door to the riches of modern quantum theory, agreement has never been reached on the conceptual foundations of this theory. The disagreement is well illustrated by the variety of opinions expressed in the other lectures we have heard in the past few days, and I shall make no attempt to summarize it. Let me turn immediately to my main purpose, which is to describe one of the most bizarre and at the same time one of the most straightforward interpretations of quantum mechanics that has ever been put forward, and that has been unjustifiably neglected since its appearance thirteen years ago.

This interpretation, which is due to Everett [1], asserts the following:

1) The mathematical formalism of quantum mechanics is sufficient as it stands. No metaphysics needs to be added to it.

2) It is unnecessary to introduce external observers or to postulate the existence of a realm where the laws of classical physics hold sway.

3) It makes sense to talk about a state vector for the whole universe.

4) This state vector never collapses, and hence the universe as a whole is rigorously deterministic.

5) The ergodic properties of laboratory measuring instruments, although strong guarantors of the internal consistency of the statistical interpretation of quantum mechanics, are inessential to its foundations.

6) The statistical interpretation itself need not be imposed *a priori*.

(*) The research behind these lectures was supported by a grant from the National Science Foundation.

In order to arrive at such assertions Everett must make certain assumptions. It is possible to be fairly precise about these without getting too technical, as they are quite simple. Basically Everett introduces two postulates, one concerning the mathematical content of the quantum formalism and one concerning the complexity of the real world:

a) *Postulate of mathematical content.* The real world, or any isolated part of it one may whish for the moment to regard as *the world*, is faithfully represented solely by the following collection of mathematical objects:

1) a vector in a Hilbert space;

2) a set of dynamical equations (derivable from a variational principle) for a set of operators which act on the Hilbert space;

3) a set of commutation relations for the operators (derived from the Poisson brackets of the classical theory by the quantization rule, in the case of those operators that possess classical analogs).

b) *Postulate of complexity.* The world is decomposable into systems and apparata.

The first postulate is a statement of the conventional mathematical apparatus of quantum physics and is hardly controversial. Or rather, it would be hardly controversial were it not for the appearance of the word «faithfully». The use of this word implies a return to naive realism and the old-fashioned idea that there can be a direct correspondence between formalism and reality. No longer, says EVERETT, are we to be bamboozled into believing that the chief issues of interpretation are epistemological rather than ontological and that the quantum realm must be viewed as a kind of ghostly world whose mathematical symbols represent potentiality rather than reality. The symbols of quantum mechanics represent reality just as much as do those of classical mechanics.

The second postulate is incomplete as I have stated it here, because the words «systems» and «apparata» have not been defined. This deficiency will be remedied by the display of actual examples in the Sections to follow.

Without drawing on any external metaphysics or mathematics other than the standard rules of logic, it is possible, from these two postulates alone, to prove the following remarkable metatheorem: ***The mathematical formalism of the quantum theory is capable of yielding its own interpretation.***

In one sense this metatheorem has already been proved. Historically, the mathematical formalism of the quantum theory was invented before its interpretation was understood. The symbols could be manipulated and certain quantities calculated, derived or guessed by a sort of magical intuition before it was known precisely what the symbols meant. It took about two years from the time of Heisenberg's first discovery for the symbols to clarify themselves.

The early history of quantum mechanics is not an isolated instance of such a situation in physics. Other examples, in which formalism came before interpretation, are to be found in the history of the Dirac wave equation, the history of quantum electrodynamics, and, more recently, the history of quantum geometrodynamics.

In all these cases, however, a certain met aphysics was present to begin with. In the early history of quantum mechanics, particularly, Bohr's metaphysical ideas played a fundamental role. Here we are trying to start from scratch and to show that much less than was previously thought is needed in the way of postulational input in order to prove the metatheorem.

Proofs in metamathematics, or metaphysics, require first the introduction of a carefully constructed syntax. If I were attempting to be rigorous I should replace words like «system», «apparatus,» «state,» «observable,» and even the statement of the metatheorem itself, by symbols subject, together with the usual mathematical symbols of the quantum formalism, to certain formal rules of manipulation but empty of any *a priori* meaning. These words would then acquire semantic content only *a posteriori*, after the consequences of Everett's postulates have been investigated.

This remains a program for the future, to be carried out by some enterprising analytical phi losopher. Here I intend to proceed quite informally, using conventional words pretty much in conventional ways. However, I shall leave them with a certain semantic vagueness at the outset. Thus I shall assume that a state is associated with a certain nonvanishing vector in Hilbert space, that an observable is associated with a certain Hermitian operator which acts on the Hilbert space, and that a dynamical entity is associated with a set of operators generating a certain algebra and satisfying certain dynamical equations, but I shall not be very precise (until later) about the nature of these associations. Precision, and hence meaning, will be acquired only by examining the quantum symbolism in a clear and specific context, namely, that of a measurement process.

## PART I.

## The Quantum Theory of Measurement.

### 1. – System, apparatus and coupling.

In its simplest form the quantum theory of measurement considers a world composed of just two dynamical entities, a *system* and an *apparatus*. It is the role of the apparatus to measure the value (a so far undefined phrase) of some system observable $s$. For this purpose the two must be coupled together (another so far undefined phrase) for a certain interval of time which we shall suppose finite. First, however, let us suppose the system and apparatus to be uncoupled so that we may examine them separately. It will be the deviation in their behavior, when coupled, from their uncoupled behavior which constitutes the measurement.

In the uncoupled condition the system is associated with a certain operator algebra, to which $s$ belongs, and the apparatus is associated with another independent operator algebra. The meaning of the word «independent,» and hence of the word «uncoupled,» is that the two algebras commute. The Hilbert space then decomposes into a Cartesian product. A choice of basis vectors reflecting this decomposition may be introduced. For example,

$$|s, A\rangle = |s\rangle|A\rangle, \tag{1.1}$$

where $s$ is an eigenvalue of $s$ and $A$ is an eigenvalue of some apparatus observable $A$, other labels that may be needed to complete the specification of the basis being here suppressed for simplicity. For later convenience we shall assume that the eigenvalue $s$ ranges over a discrete set while the eigenvalue $A$ ranges over a continuum. Conventional orthonormality and completeness conditions will then be assumed:

$$\langle s|s'\rangle = \delta_{ss'}, \qquad \langle A|A'\rangle = \delta(A - A'), \tag{1.2}$$

$$\sum_s \int |s, A\rangle\langle s, A| dA = 1. \tag{1.3}$$

The combined state of system and apparatus will be represented (in a yet to be determined sense) by a certain nonvanishing Hilbert-space vector $|\Psi\rangle$. It may happen that this vector itself decomposes into a Cartesian product

$$|\Psi\rangle = |\psi\rangle|\Phi\rangle, \tag{1.4}$$

where $|\psi\rangle$ lies in the Hilbert space generated by the operator algebra of the system and $|\Phi\rangle$ lies in the Hilbert space generated by the operator algebra of the apparatus. In this event the uncoupled system and apparatus are said to be *uncorrelated*. Each may be regarded as being in its own independent state, that of the system being represented by $|\psi\rangle$ and that of the apparatus by $|\Phi\rangle$. I shall not ask until much later (Sect. 5) how we can contrive to be sure that a real system and a real apparatus *will* be in an uncorrelated state. I shall simply assume that such a state can be produced upon demand. In fact, I shall demand it right now.

Given, then, a system and an apparatus which, when uncoupled, find themselves in an uncorrelated state, what can we say about this state when the coupling is switched on? In discussing this question I am going to take the unusual step of working in the Heisenberg picture. It is unusual because in this picture it is not possible to follow the changes of state resulting from the coupling by referring to the vector $|\Psi\rangle$, this vector being time-independent and hence given once and for all. I must instead follow the time behavior of the dynamical variables themselves, a procedure that is less familiar. The Heisenberg picture will have the merit in later Sections, however, of permitting a simple and direct comparison between the quantum and classical theories of measurement, a comparison that, while not essential in itself, will heighten the semantic content of the formalism.

When using the Heisenberg picture one should avoid careless use of conventional terminology. For example, it is not correct to refer to position or momentum as observables. Rather one must speak of the *position at a given time* or the *momentum at a given time*. More generally, observables such as $s$ or $A$ must be understood as involving particular instants or intervals of time in their intrinsic definition. Only if $s$ and $A$ commute with the energy operator do these instants become arbitrary. If $s$ and $A$ do not commute with the energy operator then neither the state associated with $|s\rangle$ nor the state associated with $|A\rangle$ can be referred to as *stationary* despite the fact that the vectors themselves are time-independent. Moreover, once a measurement of $s$ has been carried out it will generally be difficult to repeat the measurement at a later time.

In comparing the coupled and uncoupled states of the system and apparatus I shall use retarded boundary conditions. If the operator corresponding to a certain observable (of either the system or the apparatus) is constructed out of dynamical variables taken from an interval of time preceding the coupling interval, then this operator (and hence the observable itself) will remain unaffected by the coupling. Otherwise, it will generally suffer a change. In particular, the observable $s$ will generally be disturbed.

Suppose the apparatus operator $A$ is built out of dynamical variables taken from an interval of time lying to the future of the coupling interval. Then it will generally be transformed into a new operator $\bar{A}$ when the coupling is switched on. Since the coupling is not active, however, during the time interval associated with $\bar{A}$, the system and apparatus will once again during this time be dynamically independent, each running undisturbed by the other. This means that the dynamical variables out of which $\bar{A}$ is built satisfy exactly the same dynamical equations as do those out of which $A$ is built, and hence, by a well-known rule of quantum mechanics, the two sets of variables must be equivalent, *i.e.* related by a unitary transformation. In particular $A$ and $\bar{A}$ must be so related:

$$\bar{A} = \exp[-ig\mathcal{X}]A\exp[ig\mathcal{X}]\,, \tag{1.5}$$

where $\mathscr{X}$ is a certain Hermitian operator built out of the dynamical variables of *both* the system and the apparatus, and $g$ is a *coupling constant.*

The undisturbed system observable $s$, on the other hand, will not generally be related by a unitary transformation to its disturbed form $\bar{s}$. This is because the dynamical variables out of which $s$ is built belong to a time interval that generally coincides with the coupling interval, because the coupling is designed precisely to measure $s$.

The design of the coupling is governed (on paper, at any rate) by the choice of $\mathscr{X}$. We shall assume that $\mathscr{X}$ and $A$ have been chosen so as to satisfy the commutation relations

$$[\mathscr{X}, A] = is\,, \tag{1.6}$$

$$[\mathscr{X}, s] = 0\,. \tag{1.7}$$

We then find

$$\bar{A} = A + gs\,. \tag{1.8}$$

Under these conditions the coupling $g\mathscr{X}$ is said to secure a measurement of $s$, and the result of the measurement is said to be *stored* in the apparatus observable $A$, in virtue of its transformation, as a result of the coupling, into the operator $A + gs$. The observable $A$ thus constitutes a *memory unit* in the apparatus.

In Sect. 8 we shall regard the switching on of the coupling from a classical viewpoint, namely, as a modification of the combined action functional of the system and apparatus rather than as a unitary transformation. We shall see precisely how this modification must be related to the operator $\mathscr{X}$, and the question of the practical achievability of a given coupling will therefore be carried up to the point where it becomes a question of the experimenter's art rather than that of the theorist. It will be sufficient for the present simply to note that an easy way to achieve the relations (1.6) and (1.7) is to choose $\mathscr{X}$ in the form:

$$\mathscr{X} = sX\,, \tag{1.9}$$

where $X$ is the apparatus operator canonically conjugate to $A$ (*):

$$[X, A] = i\,. \tag{1.10}$$

Let me now call attention to the fact that the disturbed apparatus operator $\bar{A}$ depends on the *undisturbed* system operator $s$. This means that the apparatus records what the system observable *would have been* had there been no coupling.

---

(*) We choose units in which $\hbar = 1$.

This, is, of course, the very best kind of measurement and shows that there is in principle no limitation to the accuracy with which a single observable of a system may be determined. (Only when attempts are made to measure two observables at once are there limitations on accuracy (see Sect. 10)). This result is achieved very simply within the quantum formalism. From the classical viewpoint, however, it is not so easy to discover the modifications in the total action functional which are needed in order to compensate for the disturbance which the coupling itself produces in the system. The story of the famous Bohr-Rosenfeld paper on electromagnetic field measurements [2] is a case in point. In Sect. 8 we shall display these modifications (which are simply generalizations of Bohr's ingenious compensation devices) in detail.

In view of the fact that the apparatus records $s$ and not $\bar{s}$, we shall have little occasion to work with the latter. It is important, however, that one not become confused about this point. Given a knowledge of the explicit form of the coupling it is always possible to express the idealized undisturbed system observable $s$ in terms of the actual coupled dynamical vairables of system plus apparatus. Hence, there is nothing at all hypothetical about the operator $s$. One must only remember that its expression in terms of the coupled dynamical variables may be very complicated, so that it would, as a practical matter, generally be very difficult to find a coupling which would, in effect, reconstruct if for us and allow us to measure it a second time.

The alert student is probably becoming a little impatient at this point at my speaking of the apparatus as measuring an *operator*. No real apparatus can store an *operator* in its memory bank! What actually gets recorded is a *number*, or an analog equivalent of a number. To arrive at numbers we have to look at the state vector (1.4). With respect to the basis vectors (1.1) defined by the uncoupled system and apparatus this vector is represented by the function:

$$\langle s, A|\Psi\rangle = c_s\Phi(A)\,, \tag{1.11}$$

where

$$c_s = \langle s|\psi\rangle\,, \qquad \Phi(A) = \langle A|\Phi\rangle\,. \tag{1.12}$$

The factorization of this function into a function $c_s$ referring to the system alone and a function $\Phi(A)$ referring to the apparatus alone reflects the independence of, or lack of correlation between, the uncoupled system and apparatus. The coupled system and apparatus do not display this same independence. Relative to a basis defined by the disturbed observable $\bar{A}$, the state vector $|\Psi\rangle$ is no longer represented by a product of two functions, one of which refers only to the system and the other only to the apparatus.

The new basis is obtained by carrying out the unitary transformation (1.5).

Because

(1.13) $$s = \exp[-ig\mathscr{X}]\, s \exp[ig\mathscr{X}]\,,$$

(see eq. (1.7)), it follows that we may write

(1.14) $$|s, \bar{A}\rangle = \exp[-ig\mathscr{X}]\,|s, A\rangle\,.$$

Moreover, in view of the fact that

(1.15) $$[s, \bar{A}] = 0\,,$$

an alternative Cartesian product decomposition may be introduced:

(1.16) $$|s, \bar{A}\rangle = |s\rangle|\bar{A}\rangle\,,$$

the symbol $|\bar{A}\rangle$ denoting an eigenvector of $\bar{A}$ corresponding to the eigenvalue $\bar{A}$. Numerically the eigenvalue $\bar{A}$ is equal to the eigenvalue $A$ of $A$, and this is sometimes a source of confusion. One should take care to differentiate between $|A\rangle$ and $|\bar{A}\rangle$. They are not equal even though the numbers inside the brackets are identical. Different eigenvalues will sometimes be distinguished by adding primes, and we shall follow the rule:

(1.17) $$A' = \bar{A}'\,, \qquad A'' = \bar{A}''\,, \text{ etc.}$$

The vectors $|s, \bar{A}\rangle$ constitute a kind of mixed basis in that they are eigenvectors of a disturbed apparatus observable and an *un*disturbed system observable. This basis is nevertheless the appropriate one to use for the analysis of the measurement process, because the coupling is deliberately designed to set up a correlation between the apparatus and the value that the system observable would have assumed had there been no coupling. This correlation may be displayed by projecting the basis vectors $|s, \bar{A}\rangle$ onto the state vector $|\Psi\rangle$. From the relations

(1.18) $$\bar{A}|s, A\rangle = (A + gs)|s, A\rangle = (A + gs)|s, A\rangle = (\bar{A} + gs)|s, A\rangle\,,$$

it follows that, apart from an arbitrary phase factor that may be set equal to unity, we have

(1.19) $$|s, A\rangle = |s, \bar{A} + gs\rangle = |s\rangle|\bar{A} + gs\rangle\,,$$

or alternatively

(1.20) $$|s, \bar{A}\rangle = |s, A - gs\rangle = |s\rangle|A - gs\rangle\,,$$

where $|\bar{A}+gs\rangle$ denotes an eigenvector of $\bar{A}$ corresponding to the eigenvalue $\bar{A}+gs$ and $|A-gs\rangle$ denotes an eigenvector of $A$ corresponding to the eigenvalue $A - gs$. We therefore obtain

$$\langle s, \bar{A}|\Psi\rangle = c_s \Phi(\bar{A}-gs)\,. \tag{1.21}$$

Although the function again factorizes, one of the factors, $\Phi(\bar{A}-gs)$, now depends on the eigenvalues of *both* system and apparatus observables. The result is a correlation between system and apparatus that is easily displayed in a Figure such as Fig. 1. This Figure compares the appearance that the functions $\langle s, A|\Psi\rangle$ and $\langle s, \bar{A}|\Psi\rangle$ have, in a typical case, in the $(s, A)$ and $(s, \bar{A})$ planes respectively. What is plotted in each plane is the *effective support* of the function in question, *i.e.* the region where the function differs significantly from zero. Since the spectrum of $s$ has been taken to be discrete, these regions appear as sets of distinct vertical line segments.

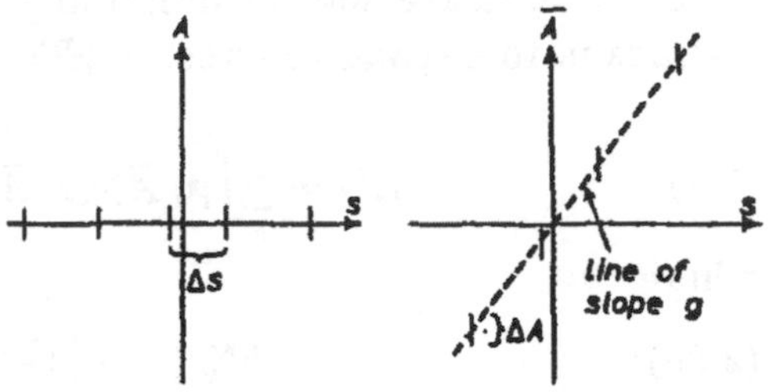

Fig. 1.

The measurement is said to be *good* if these line segments, in the case of the function $\langle s, \bar{A}|\Psi\rangle$, retain their distinctness when projected onto the $\bar{A}$ axis. In a good measurement each value of $s$ is correlated with a distinct range of values for $\bar{A}$. The quality of a measurement evidently depends not only on the choice of an appropriate apparatus and coupling but also on the state of the apparatus. The condition for a good measurement is expressed mathematically by

$$\Delta A \ll g\Delta s\,, \tag{1.22}$$

where $\Delta s$ is the minimal spacing between those eigenvalues of $s$ that are contained in the effective support of the function $c_s$, and $\Delta A$ is the root mean square deviation in $A$ from its mean value relative to the function $|\Phi(A)|^2$:

$$\Delta A^2 = \frac{\int (A-\langle A\rangle)^2|\Phi(A)|^2 \mathrm{d}A}{\int|\Phi(A)|^2\mathrm{d}A} = \langle A^2\rangle - \langle A\rangle^2\,, \tag{1.23}$$

$$\langle A^n\rangle = \frac{\langle\Psi|A^n|\Psi\rangle}{\langle\Psi|\Psi\rangle} = \frac{\langle\Phi|A^n|\Phi\rangle}{\langle\Phi|\Phi\rangle} = \frac{\int A^n|\Phi(A)|^2\mathrm{d}A}{\int|\Phi(A)|^2\mathrm{d}A}\,. \tag{1.24}$$

It will be noted that the same vector $|\Psi\rangle$ is used to represent the combined state of system and apparatus in both the coupled and uncoupled cases. This is because we are working in the Heisenberg picture and using retarded boundary conditions. Before the coupling interval the dynamical situations in the two cases are identical. The difference between these situations that occurs during and after the coupling interval is expressed by differences in the operator observables, not by differences in the state vector.

## 2. – Relative states, infinite regression, absolute chance and schizophrenia.

An alternative way of describing the correlation between system and apparatus is to expand the vector $|\Psi\rangle$ in terms of the basis vectors $|s, \bar{A}\rangle$:

$$|\Psi\rangle = \sum_s \int |s, \bar{A}\rangle\langle s, \bar{A}|\Psi\rangle \mathrm{d}\bar{A} = \sum_s c_s |s\rangle |\Phi[s]\rangle , \tag{2.1}$$

where

$$|\Phi[s]\rangle = \int |\bar{A}\rangle \Phi(\bar{A} - gs) \mathrm{d}\bar{A} , \tag{2.2a}$$

$$= \int |\bar{A} + gs\rangle \Phi(\bar{A}) \mathrm{d}\bar{A} . \tag{2.2b}$$

EVERETT [1] and WHEELER [3] have coined the expression ***relative states*** to describe the states of the apparatus represented by the vectors $|\Phi[s]\rangle$. Relative to each system state $|s\rangle$ the apparatus goes into a corresponding state $|\Phi[s]\rangle$, the bracket « [s] » denoting the fact that the value $s$ for $\boldsymbol{s}$ has been recorded in the apparatus memory. We note that when the measurement is good the relative states are orthogonal to one another:

$$\langle\Phi[s]|\Phi[s']\rangle = \delta_{ss'} \int |\Phi(A)|^2 \mathrm{d}A . \tag{2.3}$$

The decomposition (2.1) does not represent the observable $\boldsymbol{s}$ as having any unique value, unless, of course, $|\psi\rangle$ happens to be parallel to one of the basis vectors $|s\rangle$. It is here that the controversies over the interpretation of quantum mechanics start. For many people, the decomposition (2.1) does not represent an observation as having actually occurred. They conceive the apparatus to have entered a kind of schizophrenic state in which it is unable to decide what value it has found for $\boldsymbol{s}$. At the same time they cannot deny that the coupling between system and apparatus would, in the classical theory, have led to a definite outcome (see Sect. 8). They therefore try to invent schemes, both physical and metaphysical, for prodding the apparatus into making up its mind.

Many of these schemes you will hear about in the other lectures. Let me merely mention one that will *not* work, namely, the introduction of a *second* apparatus to get at the facts by looking at the first apparatus to see what it has recorded. An analysis carried out along the above lines (see also Sect. 3) quickly shows that the second performs no better than the first. It too goes into a state of schizophrenia. The same thing happens with a third apparatus, and a fourth, and so on. The schizophrenia only gets amplified by bringing in more and more of the rest of the universe; it does not disappear. This is known as the catastrophe of infinite regression.

Some have sought a resolution of the catastrophe in the fact that the unitary transformation (1.14), which is essentially due to VON NEUMANN [4], is far too simple and specialized to describe real measurements in real laboratories. Many physicists stress the fact that real apparata (including human beings) are highly ergodic systems whose initial states are imperfectly known, and that real observations may involve metastable states and nonlinear feedback loops. I shall not review these arguments here. I shall try to show that the questions raised by von Neumann's idealization of a measurement are answerable in simpler terms. One does not solve problems by making them more difficult.

The traditional way to solve the regression problem is by fiat. One asserts that after the measurement is completed (*i.e.* after the coupling interval) the state vector *collapses* to one of the elements $|s\rangle|\Phi[s]\rangle$ in the decomposition (2.1). To *which* element in the decomposition it collapses and how the collapse comes about one cannot say. One can only assign a *probability distribution* to the possible outcomes, with weights given by

$$w_s = \frac{\langle\psi|s\rangle\langle s|\psi\rangle}{\langle\psi|\psi\rangle} = \frac{|c_s|^2}{\sum_{s'}|c_{s'}|^2}\,. \tag{2.4}$$

The collapse of the state vector and the assignment of statistical weights do not follow from the dynamical equations that the dynamical variables of the system and apparatus satisfy. They are consequences of an *a priori* metaphysics that is imposed on the theory and that may be somewhat adjusted to suit convenience. For example, if one insists on adding a second apparatus, or even an indefinite string of apparata observing each other, one may leave the combined state vector uncollapsed over an arbitrary number of links in the chain, just so long as it is treated as collapsed after *some* observation somewhere along the line. In this way the regression problem is converted into a pseudoproblem.

The trouble with this solution is that physics is no longer physics; it has become metaphysics. BOHR says this in so many words: «It is wrong to think that the task of physics is to find out how nature *is*. Physics concerns what we can say about nature [5].» In a similar vein HEISENBERG remarks that the mathematics of physics «no longer describes the behavior of elementary particles, but only our knowledge of their behavior. » [6] According to this view the symbols $|\Psi\rangle$, $s$, $A$, etc., do not describe the behavior of the system and apparatus, but only a certain amount of knowledge of their behavior. As soon as a measurement is performed, knowledge is increased and the state vector collapses accordingly. The assignment of statistical weights to the elements of the decomposition (2.1) implies that these elements are to be regarded as representing the system and apparatus in exactly the same way as a statistical ensemble represents a dynamical system in classical kinetic theory. The analogy fails, however, in one crucial respect: There is always one member of the statistical ensemble that is *a priori* distinct from the others even if we do not know how to recognize it, namely, the one that is identical to the real classical system. No such element exists in the decomposition (2.1); reality, for quantum systems, dissolves into a metaphysical mirage.

The traditional interpretation of quantum mechanics evidently confuses two concepts that ought really to be kept distinct, namely, probability as it relates to quantum mechanics and probability as it is understood in statistical mechanics. Quantum mechanics is a theory that attempts to describe in mathematical language a world in which chance is not a measure of our ignorance but is *absolute*. It is inevitable that it lead to descriptions of the world by vectors like (2.1) which contain, in a single superposition, all the possible outcomes of a given measurement. However, precisely because quantum-mechanical chance is *not* a measure of our ignorance, we ought not to tamper with the state vector merely because we acquire new information as a result of a measurement.

The obstacle to taking such an honest view of things, of course, is that it forces us to believe in the reality of all the simultaneous worlds represented in the superposition (2.1), in each of which the measurement has yielded a different outcome. If we are to follow EVERETT in letting the formalism tell its own story, without benefit of *a priori* metaphysics, then we must be willing to admit that even the entire universe may be faithfully represented by (one might even say isomorphic to) a superposition like (2.1) but of vastly greater complexity. Our universe must be viewed as constantly splitting into a stupendous number of branches, all resulting from the measurementlike interactions between its myriads of components. Because there exists neither a mechanism within the framework of the formalism nor, by definition, an entity outside of the universe that can designate which branch of the grand superposition is the « real » world, all branches must be regarded as equally real.

To see what this multi-world concept implies one need merely note that because every cause, however microscopic, may ultimately propagate its effects throughout the universe, it follows that every quantum transition taking place on every star, in every galaxy, in every remote corner of the universe is splitting our local world on earth into myriads of copies of itself. Here is schizophrenia with a vengeance! The idea of $10^{100+}$ slightly different copies of oneself all constantly splitting into further copies, which ultimately become unrecognizable, is hard to reconcile with the testimony of our senses, namely, that we simply do not split. EVERETT [1] compares this testimony with that of the anti-Copernicans in the time of GALILEO, who did not feel the earth move. We know now that Newtonian gravitational theory, within the framework of classical physics, accounts completely for this lack of sensation. The present difficulty has a similar solution. We shall show, in the next Section, that to the extent to which we can be regarded simply as automata, and hence on a par with ordinary measuring apparata, *the laws of quantum mechanics do not allow us to feel ourselves split.*

## 3. – Unobservability of the splits.

Let us begin by asking what would happen, in the case of the measurement described by the superposition (2.1), if we introduced a second apparatus that not only looks at the memory bank of the first apparatus but also carries out an independent direct check on the value of the system observable. If the splitting of the universe is to be unobservable the results had better agree.

For the system we again introduce the basis vector $|s\rangle$, and for the apparata in the uncoupled state we introduce basis vectors $|A_1\rangle$ and $|A_2, B_2\rangle$ respectively. The total measurement will be carried out in two steps. In the first, both apparata observe the system observable $s$, by means of couplings $g\mathscr{X}_1$ and $g\mathscr{X}_2$ that satisfy

$$(3.1)\qquad [\mathscr{X}_1, A_1] = [\mathscr{X}_2, A_2] = is\,,$$

$$(3.2)\qquad [\mathscr{X}_1, s] = [\mathscr{X}_2, s] = 0\,,$$

$$(3.3)\qquad [\mathscr{X}_1, A_2] = [\mathscr{X}_1, B_2] = [\mathscr{X}_2, A_1] = [\mathscr{X}_2, B_2] = 0\,.$$

The unitary transformation generated by this coupling is

$$(3.4)\qquad \exp[-ig(\mathscr{X}_1 + \mathscr{X}_2)](|s\rangle|A_1\rangle|A_2, B_2\rangle) = \\ = |s\rangle|\bar{A}_1\rangle|\bar{A}_2, B_2\rangle = |s\rangle|A_1 - gs\rangle|A_2 - gs, B_2\rangle\,.$$

In the second step apparatus 2 reads the memory contents of apparatus 1 by means of a coupling $\bar{g}\mathscr{Y}$ that satisfies

$$(3.5)\qquad [\mathscr{Y}, B_2] = iA_1\,,\qquad [\mathscr{Y}, A_1] = 0\,,$$

$$(3.6)\qquad [\mathscr{Y}, s] = 0\,,\qquad [\mathscr{Y}, A_2] = 0\,.$$

In the absence of step 1 the unitary transformation generated by this coupling would be

$$(3.7)\qquad \exp[-i\bar{g}\mathscr{Y}](|s\rangle|A_1\rangle|A_2, B_2\rangle) = |s\rangle|A_1\rangle|A_2, \bar{B}_2\rangle = |s\rangle|A_1\rangle|A_2, B_2-\bar{g}A_1\rangle\,.$$

When both couplings are introduced consecutively the undisturbed observables $A_1, A_2, B_2$ get transformed into disturbed observables (*) $\bar{A}_1, \bar{A}_2, \bar{B}_2$, and the basis vectors defined by the latter are obtained *via* the overall unitary transformation:

$$(3.8)\qquad \exp[-i\bar{g}\bar{\mathscr{Y}}]\exp[-ig(\mathscr{X}_1+\mathscr{X}_2)](|s\rangle|A_1\rangle|A_2, B_2\rangle) =$$
$$= \exp[-ig(\mathscr{X}_1+\mathscr{X}_2)]\exp[-i\bar{g}\mathscr{Y}](|s\rangle|A_1\rangle|A_2, B_2\rangle) =$$
$$= |s\rangle|A_1-gs\rangle|A_2-gs, B_2-\bar{g}A_1\rangle\,,$$

where $\bar{\mathscr{Y}}$ is the operator into which $\mathscr{Y}$ is transformed by the coupling of step 1.

If the system and apparata are initially uncorrelated, so that the combined state vector has the form

$$(3.9)\qquad |\Psi\rangle = |\psi\rangle|\Phi_1\rangle|\Phi_2\rangle\,,$$

then the projection of this vector onto the undisturbed basis has the form

$$(3.10)\qquad \langle s, A_1, A_2, B_2|\Psi\rangle = c_s\Phi_1(A_1)\Phi_2(A_2, B_2)\,,$$

where

$$(3.11)\qquad c_s = \langle s|\psi\rangle\,,\quad \Phi_1(A_1) = \langle A_1|\Phi_1\rangle\,,\quad \Phi_2(A_2, B_2) = \langle A_2, B_2|\Phi_2\rangle\,.$$

Its projection onto the disturbed basis, however, has the form

$$(3.12)\qquad \langle s, \bar{A}_1, \bar{A}_2, \bar{B}_2|\Psi\rangle = c_s\Phi_1(\bar{A}_1-gs)\Phi_2(\bar{A}_2-gs, \bar{B}_2-\bar{g}\bar{A}_1)\,,$$

which yields the decomposition

$$(3.13)\qquad |\Psi\rangle = \sum_s\int d\bar{A}_1\int d\bar{A}_2\int d\bar{B}_2 c_s|s\rangle|\bar{A}_1\rangle|\bar{A}_2, \bar{B}_2\rangle\cdot$$
$$\cdot\Phi_1(\bar{A}_1-gs)\Phi_2(\bar{A}_2-gs, \bar{B}_2-\bar{g}\bar{A}_1)$$
$$= \sum_s\int d\bar{A}_1\int d\bar{A}_2\int d\bar{B}_2 c_s|s\rangle|\bar{A}_1+gs\rangle|\bar{A}_2+gs, \bar{B}_2+\bar{g}(\bar{A}_1+gs)\rangle\Phi_1(\bar{A}_1)\Phi_2(\bar{A}_2, \bar{B}_2).$$

(*) The observation of $\bar{A}_1$ by the second apparatus (via the coupling $\bar{g}\mathscr{Y}$) may further disturb this observable, but this is unimportant in the present argument.

In order to keep the interpretation of this decomposition as simple as possible it is convenient to assume that the effective support of the function $\Phi_1(A_1)$ is very narrow, so that the measurement of $\bar{A}_1$ by apparatus 2 resembles the measurement of an observable having a discrete spectrum. Explicitly we require

$$\Delta A_1 \ll \Delta B_2/\bar{g}\,, \tag{3.14}$$

where $\Delta A_1$ and $\Delta B_2$ are defined in an obvious manner. When (3.14) holds we may approximate (3.13) by a simple decomposition into relative states:

$$|\Psi\rangle = \sum_s c_s|s\rangle\, |\Phi_1[s]\rangle |\Phi_2[s, \langle A_1\rangle + gs]\rangle \tag{3.15}$$

with

$$|\Phi_1[s]\rangle \quad = \int |\bar{A}_1 + gs\rangle \Phi_1(\bar{A}_1)\mathrm{d}\bar{A}_1 \tag{3.16}$$

$$|\Phi_2[s, A_1]\rangle = \int \mathrm{d}\bar{A}_2 \int \mathrm{d}\bar{B}_2 |\bar{A}_2 + gs, \bar{B}_2 + \bar{g}A_1\rangle \Phi(\bar{A}_2, \bar{B}_2)\,. \tag{3.17}$$

The couplings will have yielded good measurements if, in addition to (3.14), we have

$$\Delta A_1 \ll g\Delta s\,, \qquad \Delta A_2 \ll g\Delta s\,, \qquad \Delta B_2 \ll \bar{g}g\Delta s\,, \tag{3.18}$$

so that the relative states become orthogonal:

$$\langle \Phi_1[s]|\Phi_1[s']\rangle = \delta_{ss'} \int |\Phi_1(A_1)|^2 \mathrm{d}A_1\,, \tag{3.19}$$

$$\langle \Phi_2[s, \langle A_1\rangle + gs]|\Phi_2[s', \langle A_1\rangle + gs']\rangle = \delta_{ss'} \int \mathrm{d}A_2 \int \mathrm{d}B_2 |\Phi_2(A_2, B_2)|^2\,. \tag{3.20}$$

The combined state vector $|\Psi\rangle$ is again revealed (eq. (3.15)) as a linear superposition of vectors, each of which represents the system observable $s$ as having asssumed one of its possible values. Although the value varies from one element of the superposition to another, not only do both apparata within a given element observe the value appropriate to that element, but also, by straightforward communication, they agree that the results of their observations are identical. Apparatus 2 may be assumed to have « known in advance » that the « mean value » of $A_1$ was $\langle A_1\rangle$. When, after the coupling $\bar{g}\bar{\mathscr{Y}}$, apparatus 2 « sees » that this mean value has shifted to $\langle A_1\rangle + gs$, it then « knows » that apparatus 1 has obtained the same value for $s$, namely $s$, as it did.

It is not difficult to devise increasingly complicated situations, in which, for example, the apparata can make decisions by switching on various couplings depending *on the outcome of other observations.* No inconsistencies will ever arise, however, that will permit a given apparatus to be aware of more than one world at a time. Not only is its own memory content always self-consistent (think of the two apparata above as a single apparatus which can communicate with itself) but consistency is always maintained as well in rational discourse with other automata. Extending these conclusions to the universe as a whole we see that from the point of view of any automaton, within any branch of the universal state vector, schizophrenia cannot be blamed on quantum mechanics. We also see that the catastrophe of infinite regres*sion is not a catastrophe at all.*

This is a good place to take stock of what the formalism has taught us so far about the meaning of the symbols appearing in it:

1) An apparatus that measures an observable never records anything but an eigenvalue of the corresponding operator, at least if the measurement is good.

2) *The operator corresponding to a given observable represents not the* value of the observable, but rather *all* the values that the observable can assume under various conditions, the values themselves being the eigenvalues.

3) The dynamical variables of a system, being operators, do not represent the system other than generically. That is, they represent not the system as it really is, but rather all the situations in which the system might conceivably find itself.

4) Which situation a system is actually in is specified by the state vector. Reality is therefore described jointly by the dynamical variables and the state vector. This reality is not the reality we customarily think of, but is a reality composed of many worlds.

This list is unfortunately not yet sufficient to tell us how to apply the formalism to practical problems. The symbols that describe a given system, namely, the state vector and the dynamical variables, describe not only the system as it is observed in one of the many worlds comprising reality, but also the system as it is seen in all the other worlds. We, who inhabit only one of these worlds, have no symbols to describe our world alone. Because we have no access to the other worlds it follows that we are unable to make rigorous predictions about reality as we observe it. Although reality as a whole is completely deterministic, our own little corner of it suffers from indeterminism. The interpretation of the quantum mechanical formalism (and hence the proof of Everett's metatheorem) is complete only when we show that this indeterminism is nevertheless limited by rigorous statistical laws.

**4. – The statistical interpretation of quantum mechanics.**

When the apparatus of Sect. 1 measures the system observable $s$ we cannot predict what value it will record, except that this value will be an eigenvalue lying in the support of the function $c_s$. Suppose, however, that the apparatus makes repeated measurements on an ensemble of uncorrelated identical systems that are initially in identical states. The total state vector then has the form

$$|\Psi\rangle = |\psi_1\rangle|\psi_2\rangle \dots |\Phi\rangle , \tag{4.1}$$

where

$$\langle s_n|\psi_n\rangle = c_{s_n} \text{ for all } n , \tag{4.2}$$

and the successive measurements are described by the action of unitary operators on a set of basis vectors

$$|s_1\rangle|s_2\rangle \dots |A_1, A_2 \dots\rangle , \tag{4.3}$$

appropriate to the no-coupling situation. If the apparatus observes each system exactly once, in sequence, then the $n$-th measurement is described by a unitary transformation of the form

$$\exp[-ig\mathscr{X}_n](|s_1\rangle|s_2\rangle \dots |A_1, A_2, \dots, A_n, \dots\rangle) = \\ = |s_1\rangle|s_2\rangle \dots |A_1, A_2, \dots, A_n - gs_n, \dots\rangle \tag{4.4}$$

where the coupling $g\mathscr{X}_n$ satisfies

$$[\mathscr{X}_n, A_m] = i\delta_{nm}s_n , \qquad [\mathscr{X}_n, s_m] = 0 . \tag{4.5}$$

After $N$ measurements have taken place, the first $N$ of the undisturbed apparatus observables $A_1, A_2, \dots,$ find themselves transformed into disturbed variables $\bar{A}_1 \dots \bar{A}_N$ given by

$$\bar{A}_n = \exp[-ig\mathscr{X}_n]A_n \exp[ig\mathscr{X}_n] , \tag{4.6}$$

and the basis vectors defined by the disturbed variables are

$$|s_1\rangle|s_2\rangle \dots |\bar{A}_1 \dots \bar{A}_N, A_{N+1} \dots\rangle = |s_1\rangle|s_2\rangle) \dots |A_1 - gs_1, \dots, A_N - gs_N, A_{N+1} \dots\rangle . \tag{4.7}$$

If we now decompose $|\Psi\rangle$ in terms of the basis vectors (4.7) we find

$$|\Psi\rangle = \sum_{s_1, s_2 \dots} c_{s_1}c_{s_2} \dots |s_1\rangle|s_2\rangle \dots |\Phi[s_1 \dots s_N]\rangle , \tag{4.8}$$

where

$$(4.9)\qquad |\Phi[s_1 \dots s_N]\rangle$$
$$= \int d\bar{A}_1 \dots \int dA_{N+1} \dots |\bar{A}_1 + gs_1, \dots, \bar{A}_N + gs_N, A_{N+1} \dots\rangle \Phi(\bar{A}_1 \dots \bar{A}_N, A_{N+1} \dots),$$

$$(4.10)\qquad \Phi(A_1, A_2 \dots) = \langle A_1, A_2 \dots |\Phi\rangle .$$

It will be observed that although every system is initially in exactly the same state as every other, the apparatus, as represented by the relative state vectors $|\Phi[s_1 \dots s_N]\rangle$, does not generally record a sequence of identical values for the system observable, even within a single element of the superposition (4.8). Each *memory sequence* $s_1 \dots s_N$ yields a distribution of possible values for the system observable.

Each of these distributions may be subjected to a statistical analysis. The first and simplest part of such an analysis is the calculation of the *relative frequency function* of the distribution:

$$(4.11)\qquad f(s; s_1 \dots s_N) = \frac{1}{N}\sum_{n=1}^{N} \delta_{ss_n} .$$

We shall need the following easily verified properties of this function (see Appendix A):

$$(4.12)\qquad \sum_{s_1 \dots s_N} f(s; s_1 \dots s_N) w_{s_1} \dots w_{s_N} = w_s ,$$

$$(4.13)\qquad \sum_{s_1 \dots s_N} [f(s; s_1 \dots s_N) - w_s]^2 w_{s_1} \dots w_{s_N} = \frac{1}{N} w_s(1 - w_s),$$

where the $w$'s are any numbers that, taken all together, add up to unity.

Let us choose for the $w$'s the numbers defined in eq. (2.4), and let us introduce the function

$$(4.14)\qquad \delta(s_1 \dots s_N) = \sum_s [f(s; s_1 \dots s_N) - w_s]^2 .$$

This is the first of a hierachy of functions that measure the degree to which the sequence $s_1 \dots s_N$ deviates from a random sequence with weights $w_s$. Let $\varepsilon$ be an arbitrarily small positive number. We shall call the sequence $s_1 \dots s_N$ *first-random* if $\delta(s_1 \dots s_N) < \varepsilon$ and *nonfirst-random* otherwise.

Suppose now we remove from the superposition (4.8) all those elements for which the apparatus memory sequence is nonfirst-random. Denote the result by $|\Psi_N^\varepsilon\rangle$ and define

$$(4.15)\qquad |\chi_N^\varepsilon\rangle = |\Psi\rangle - |\Psi_N^\varepsilon\rangle = \sum_{\substack{s_1, s_2 \dots \\ \delta(s_1 \dots s_N) \geq \varepsilon}} c_{s_1} c_{s_2} \dots |s_1\rangle |s_2\rangle \dots |\Phi[s_1 \dots s_N]\rangle .$$

Then, making use of (4.13), we find

$$\langle\chi_N^\varepsilon|\chi_N^\varepsilon\rangle = \langle\Psi|\Psi\rangle \sum_{\substack{s_1\ldots\\ \delta(s_1\ldots s_N)\geq\varepsilon}} w_{s_1}w_{s_2}\ldots = \langle\Psi|\Psi\rangle \sum_{\substack{s_1\ldots s_N\\ \delta(s_1\ldots s_N)\geq\varepsilon}} w_{s_1}\ldots w_{s_N} \tag{4.16}$$

$$< \frac{1}{\varepsilon}\langle\Psi|\Psi\rangle \sum_{s_1\ldots s_N} \delta(s_1\ldots s_N)\, w_{s_1}\ldots w_{s_N} = \frac{1}{N\varepsilon}\langle\Psi|\Psi\rangle \sum_s w_s(1-w_s) < \frac{1}{N\varepsilon}\langle\Psi|\Psi\rangle\,.$$

From this it follows that no matter how small we choose $\varepsilon$ we can always find an $N$ big enough so that the norm of $|\chi_N^\varepsilon\rangle$ becomes smaller than any positive number. This means that

$$\lim_{N\to\infty} |\Psi_N^\varepsilon\rangle = |\Psi\rangle\,. \tag{4.17}$$

It will be noted that, because of the orthogonality of the basis vectors $|s_1\rangle|s_2\rangle\ldots$, this result holds regardless of the quality of the measurements, *i.e.* independently of whether or not the condition

$$\langle\Phi[s_1\ldots s_N]|\Phi[s_1'\ldots s_N']\rangle = \langle\Phi|\Phi\rangle \prod_{n=1}^{N} \delta_{s_n s_n'} \tag{4.18}$$

for good measurements is satisfied.

A similar result is obtained if $|\Psi_N^\varepsilon\rangle$ is redefined by excluding, in addition, elements of the superposition (4.2) whose memory sequences fail to meet any finite combination of the infinity of other requirements for a random sequence. Moreover, no other choice for the $w$'s but (2.4) will work. *The conventional statistical interpretation of quantum mechanics thus emerges from the formalism itself.* Nonrandom memory sequences in the superposition (4.8) are of *measure zero* in the Hilbert space, in the limit $N\to\infty$ (*). Each automaton (that is, apparatus *cum* memory sequence) in the superposition sees the world obey the familiar statistical quantum laws. This conclusion obviously admits of immediate extension to the world of cosmology. Its state vector is like a tree with an enormous number of branches. Each branch corresponds to a possible universe-as-we-actually-see-it.

The alert student may now object that the above argument contains an element of circularity. In order to derive the *physical* probability interpreta-

(*) Everett's original derivation of this result [1] invokes the formal equivalence of measure theory and probability theory, and is rather too brief to be entirely satisfying. The present derivation is essentially due to GRAHAM [7] (see also ref. [8]). A more rigorous treatment of the statistical interpretation question, which deals carefully with the problem of defining the Hilbert space in the limit $N\to\infty$, has been given by HARTLE [9].

tion of the numbers $w_s$, based on sequences of observations, we have introduced a *nonphysical* probability concept, namely that of the measure of a subspace in Hilbert space. The latter concept is alien to experimental physics because it involves many elements of the superposition at once, and hence many simultaneous worlds, which are supposed to be unaware of one another.

The problem that this objection raises is like many that have arisen in the long history of probability theory. It should be stressed that no element of the superposition is, in the end, excluded. All the worlds are there, even those in which everything goes wrong and all the statistical laws break down. The situation is similar to that which we face in ordinary statistical mechanics. If the initial conditions were right the universe-as-we-see-it *could* be a place in which heat sometimes flows from cold bodies to hot. We can perhaps argue that in those branches in which the universe makes a habit of misbehaving in this way, life fails to evolve, so no intelligent automata are around to be amazed by it (*).

## 5. – Remaining questions.

The arguments of the preceding Section complete the (informal) proof of Everett's metatheorem and, incidentally, provide the information that physical states are to be identified with *rays* in Hilbert space rather than with the vectors themselves. This follows from the fact that no observable consequences of any of the couplings we have introduced are changed if we multiply the vectors $|\psi\rangle$, $|\psi_n\rangle$, $|\Phi\rangle$, etc. by arbitrary nonvanishing complex numbers. The statistical weights $w_s$, in particular, are unaffected thereby.

There remain, however, a number of questions that need to be cleared up. The first is a practical one. How does it happen that we are able, without running into inconsistencies, to include as much or as little as we like of the real world of cosmology in the state vectors and operators we use? Why should we be so fortunate as to be able, in practice, to avoid dealing with the state vector and operator algebra of the whole universe?

The answer to this question is to be found in the statistical implications of sequences of measurements of the kind considered in Sect. 4. Consider one of the memory sequences in the superposition (4.8). This memory sequence defines an average value for the numbers $s_1 \dots s_N$, given by

$$\langle s\rangle_{s_1\dots s_N} = \sum_s s f(s;\, s_1 \dots s_N)\,. \tag{5.1}$$

(*) It may also happen that the arrow of time is reversed for some of the branches. This would be the case if the state vector of the universe were invariant under time reversal.

If the sequence is random, as it is increasingly likely to be as $N$ becomes large, this average will differ only by an amount of order $\varepsilon$ from the prototypical average:

$$\langle s \rangle = \sum_s s w_s \,, \tag{5.2}$$

This latter average may be regarded as an *expectation value* for the observable $s$ of a *single* system in the state $|\psi\rangle$.

This expectation value may be expressed in the alternative forms

$$\langle s \rangle = \sum_s \frac{\langle \psi | s \rangle s \langle s | \psi \rangle}{\langle \psi | \psi \rangle} = \frac{\langle \psi | s | \psi \rangle}{\langle \psi | \psi \rangle} \,. \tag{5.3}$$

In the second of these forms the basis vectors $|s\rangle$ do not appear. It is evident therefore that had we chosen to introduce a different apparatus, designed to measure some observable $r$ not equal to $s$, a sequence of repeated measurements would have yielded in this case an average approximately equal to

$$\langle r \rangle = \frac{\langle \psi | r | \psi \rangle}{\langle \psi | \psi \rangle} \,, \tag{5.4}$$

in which again no basis vectors appear. We can, of course, if we like, reintroduce the basis vectors $|s\rangle$, obtaining

$$\langle r \rangle = \langle \psi | \psi \rangle^{-1} \sum_{s,s'} c_s^* \langle s | r | s' \rangle c_{s'} \,. \tag{5.5}$$

Now suppose that instead of performing a sequence of identical measurements to obtain an experimental value for $\langle r \rangle$, we first measure $s$ in each case and *then* perform a statistical analysis on $r$. This could be accomplished by introducing a second apparatus which performs a sequence of observations on a set of identical two-component systems all in identical states given by the vector $|\Psi\rangle$ of eq. (2.1). Each of the latter systems is composed of one of the original systems together with an apparatus that has just measured the observable $s$. The job of the second apparatus is to make observations of the $r$'s ($r_1$, $r_2$, etc.) of these two-component systems. Because a measurement of the corresponding $s$ has already been carried out in each case, however, these $r$'s are not the undisturbed $r$'s but the $r$'s resulting from the couplings $g\mathscr{X}$. That is, what we are really measuring in every case is the observable

$$\bar{r} = \exp\left[-ig\mathscr{X}\right] r \exp\left[ig\mathscr{X}\right] . \tag{5.6}$$

Within each element of the grand superpostion the second apparatus will obtain a sequence $\bar{r}_1 \dots \bar{r}_N$ of values for $\bar{r}$. If the element is typical the average of this sequence will, by analogy with our previous analysis, be approximately equal to

$$(5.7) \qquad \langle\bar{r}\rangle_s = \frac{\langle\Psi|\bar{r}|\Psi\rangle}{\langle\Psi|\Psi\rangle} =$$

$$= \langle\Psi|\Psi\rangle^{-1} \sum_{s,s'} \int d\bar{A} \int d\bar{A}' \langle\Psi|s,\bar{A}\rangle\langle s,\bar{A}|\exp[-ig\mathscr{X}]r\exp[ig\mathscr{X}]|s',\bar{A}'\rangle\langle s',\bar{A}'|\Psi\rangle =$$

$$= \langle\Psi|\Psi\rangle^{-1} \sum_{s,s'} \int d\bar{A} \int d\bar{A}' \langle\Psi|s,\bar{A}\rangle\langle s,A|r|s',A'\rangle\langle s',\bar{A}'|\Psi\rangle =$$

$$= \langle\psi|\psi\rangle^{-1}\langle\Phi|\Phi\rangle^{-1} \sum_{s,s'} \int c_s^* \langle s|r|s'\rangle c_s \Phi^*(\bar{A}-gs)\Phi(\bar{A}-gs')d\bar{A} .$$

If the measurements of $s$ are good in every case, so that the relative states (2.2) satisfy the orthogonality property (2.3), then this average reduces to

$$(5.8) \qquad \langle\bar{r}\rangle_s = \sum_s w_s \langle s|r|s\rangle = \mathrm{Tr}(\rho_s r) ,$$

where $\rho_s$ is the *density operator*:

$$(5.9) \qquad \rho_s = \sum_s |s\rangle w_s \langle s| .$$

The averages (5.5) and (5.8) are generally not equal. In (5.8) the measurement of $s$, which the first apparatus has performed, has destroyed the quantum interference effects that are still present in (5.5). This means that the elements of the superposition (2.1) may, insofar as the subsequent *quantum* behavior of the system is concerned, be treated *as if* they were members of a statistical ensemble. This is what allows us, *in practice*, to follow the traditional prescription of collapsing the state vector after a measurement has occurred, and to use the techniques of ordinary statistical mechanics, in which we change the description of the state upon receipt of new information. It is also what permits us to introduce, and study the quantum behavior of, systems having well-defined initial states, without at the same time introducing into the mathematical formalism the apparata which prepared the systems in those states. In brief, it is what allows us to start at any point in any branch of the universal state vector without worrying about previous or simultaneous branches, and to focus our attention on the dynamical systems of immediate relevance without worrying about the rest of the universe.

It is, of course, possible in principle (although virtually impossible in practice) to restore the interference effects of eq. (5.5) by reversing the coupling (*i.e.* $\mathscr{X} \to -\mathscr{X}$) so that the relative state vectors $|\Phi[s]\rangle$ are no longer orthogonal. But then the correlations between system and apparatus are destroyed, the apparatus memory is wiped out, and no measurement results. If one attempts to maintain the correlations by sneaking in a second apparatus to « have a look » before the interference effects are restored, then the mathematical description of the situation must be amplified to include the second apparatus, and the orthogonality of *its* relative state vectors will destroy the interference effects.

The overwhelming impracticability of restoring interference effects by reversing a measurement has been stressed by DANERI, LOINGER and PROSPERI [10] in a well-known study of the critical features, such as metastability, which characterize typical measuring apparata. One should be careful not to conclude from their study, however, that the state vector *really* collapses as the traditionalists claim. Although it is true that the ergodic properties of apparata having many degrees of freedom, which DANERI *et al.* describe, greatly expedite the orthogonalization of the relative state vectors, the interference effects are *in principle* still retrievable. LOINGER states this explicitly in a later paper [11]. In referring to his work with DANERI and PROSPERI, he says that the interference effects are only « *practically* absent, » and adds the following comment: « We did not assert that superpositions of vectors corresponding to different macroscopic states are impossible. Indeed, this possibility is firmly rooted in the formal structure of quantum theory and cannot be eliminated. » LOINGER is here, wittingly or unwittingly, casting his vote for Everett's multi-world concept, despite the fact that his papers with DANERI and PROSPERI claim to support traditionalist doctrine.

Although the ergodic properties of measuring instruments cannot be used to prove that the state vector collapses, they may very likely be of help in answering two questions of a somewhat different nature: 1) How can we be sure that uncorrelated system-apparatus states can be produced upon demand, as was assumed in Sect. 1? 2) Why is it so easy in practice to make good measurements?

The first question may be answered formally as follows: We simply prepare both system and apparatus in the state required, taking care that they remain uncoupled to one another during this process. Because, as we have already remarked, the devices that effect the preparations need not themselves be included in the state vector, the states of system and apparatus will *ipso facto* be independent, and their combined state will be uncorrelated. A preparation, however, is just a special case of a good measurement. In order to answer question 1) properly, therefore we have first to answer question 2).

Question 2) may be rephrased thus: Why is it so easy to find apparata in states which satisfy the condition (1.22) for good measurements? In the case of macroscopic apparata it is well known that a small value for the mean square deviation of a macroscopic observable is a fairly stable property of the apparatus. But how does the mean square deviation become so small in the first place? Why is a large value for the mean square deviation of a macroscopic observable virtually never, in fact, encountered in practice?

It is likely that the ergodic properties of a macroscopic apparatus bring about an automatic condensation of almost every initial state of the apparatus into a superposition of orthogonal states each of which *a*) corresponds to a distinct world-branch that will (almost) never in the future interfere with any other branch, and *b*) satisfies a narrowness condition like (1.22) relative to every macroscopic observable $A$. However, a proof of this does not yet exist. It remains a program for the future.

## PART II.

## The Classical Description of the Measurement Process.

### 6. – Action functional, dynamical equations and small disturbances.

The history of quantum mechanics would undoubtedly be far different from what it is were it not for the fact that many quantum systems have classical analogs. The planets of Newtonian celestial mechanics are analogs of the elementary particles, and the classical electromagnetic field of Maxwell is the prototypical analog of the abstract quantum fields that one often introduces to describe bosons. Even the anticommuting fields used to describe fermions are only one formal step removed from the fields of classical physics. It is hard to decide whether the development of quantum physics would have been easier or harder had there not existed the wealth of systems that can be treated very accurately purely by classical physics and that have served for centuries to condition our minds in ways that cause all of us some pain in making the readjustment to quantum ways of thinking.

In any event, the practical existence of a classical realm has compensated to a large extent for the prejudices that it has ingrained in us. The single most powerful tool that made the development of quantum mechanics possible was the correspondence principle invented by BOHR. It is difficult to imagine how the modern theory could have been discovered without this tool. When we use the same symbols to describe the position and energy of an elementary particle as we do to describe the position and energy of a planet, we are demonstrating both the usefulness and the validity of the correspondence principle in its most general (modern) form. The mental images we attach to the symbols are basically classical whether the symbols are $q$-numbers or $c$-numbers. Since such symbols *have* to be introduced when we want to get away from the abstract generalities of the quantum formalism and study the structural detail of actual dynamical systems, it follows that structral details are almost always envisaged classically. Even such abstract structures as the $V-A$ coupling of β-decay theory are colloquially described in terms of the annihilation and creation of *particles* having definite *helicities*, *i.e.*, in classical language.

It is always good to try to construct explicit models to illustrate abstract concepts. This second part of my lectures is aimed at revealing some of the details about the couplings between systems and apparata that were omitted in Part. I. Since I want my remarks to apply to very general classes of systems, if not all, the discussion will remain somewhat formal and schematic. However, since structural details of a general type now become important the language used will be mainly classical.

In quantum mechanics it is often more convenient to study the Schrödinger equation satisfied by the unitary operator that effects displacements in time than it is to study the dynamical equations satisfied by the operator observables themselves. In classical mechanics, however, the dynamical equations move to center stage. The most important property of the dynamical equations is that they may always be derived from a variational principle, based on an *action functional* which may be regarded as summarizing, in one compact expression, all the dynamical properties of the system under consideration. In fact, one will not go far astray in simply identifying a system with its action functional.

I shall denote the action functional of a system by the letter « $S$ ». $S$ is a function of the values assumed, over a finite open but otherwise arbitrary time interval, by a set of functions $q^i(t)$ that describe the dynamical trajectory of the system. The $q^i$ are known as the *dynamical variables*, and the *dynamical equations* are the functional differential equations

$$\frac{\delta S}{\delta q^i(t)} = 0 \,. \tag{6.1a}$$

(The time interval involved in the definition of $S$ will be assumed to embrace all instants at which it may be desired to perform functional differentiations.) If $S$ is expressible as the time integral of a Lagrangian function then the dynamical equations are ordinary differential equations.

In field theories the index $i$ becomes a continuous index, representing points in space as well as field components. It will be a convenient abbreviation to absorb the time label $t$, as well, into the index $i$. The summation convention over repeated indices, which I shall adopt, will then be understood to include integrations over space and/or time. The symbols in eq. (6.1*a*) will be replaced by

$$S_{,i} = 0\,, \tag{6.1b}$$

where the comma followed by an index denotes functional differentiation. Repeated functional differentiation will be indicated by the addition of further indices. It should be noted that functional differentiation, like ordinary differentiation, is commutative (*). In particular, we have

$$S_{,ij} = S_{,ji}\,. \tag{6.2}$$

In the quantum theory of the system $S$ the same dynamical equations continue to hold as in the classical theory, with the symbols being understood as operators (a transition I shall generally make explicit through the replacement of lightface symbols by boldface ones), provided appropriate care is taken in ordering the operators and provided, in the field theory case, certain renormalization constants are introduced. The equations also hold, approximately, in the quasi-classical limit, in the case of systems having a finite number of degrees of freedom, if the symbol $q^i$ is understood as standing for the expectation value $\langle q^i \rangle$ of the corresponding quantum-dynamical variable.

Suppose now we choose some solution $q^i$ of the dynamical eqs. (6.1), and suppose we ask how this solution would be modified if the action were given an infinitesimal increment proportional to some observable $r$:

$$S \to S + \varepsilon r\,. \tag{6.3}$$

The answer will, of course, depend on the boundary conditions we adopt. Let us choose retarded boundary conditions. Then the solution is unchanged prior to the time interval involved in the definition of $r$ but acquires an infinitesimal increment $\delta q^i$ thereafter. The new solution $q^i + \delta q^i$ satisfies the equation:

$$0 = S_{,i}[q + \delta q] + \varepsilon r_{,i}[q + \delta q] = S_{,ij}\delta q^j + \varepsilon r_{,i}\,, \tag{6.4}$$

correct to first infinitesimal order. This is an inhomogeneous linear equation in $\delta q^i$ known as the *equation of small disturbances*. The solution that incorporates the stated boundary conditions is

$$\delta q^i = \varepsilon G^{-ij} r_{,j}\,, \tag{6.5}$$

(*) A similar formalism can be set up for anticommuting variables, in which functional differentiation is anticommutative.

where $G^-$ is the retarded Green's function of $S_{,ij}$:

$$S_{,ik}G^{-kj} = -\delta_i{}^j \,, \tag{6.6}$$

$$G^{-ij} = 0 \qquad \text{when } t_i < t_j \,, \tag{6.7}$$

$t_i$ being the instant of time labeled by the index $i$.

In Lagrangian theories $S_{,ij}$ is essentially a differential operator, and in specific cases approximate, if not exact, expressions can often be obtained for its Green's functions. The commutativity condition (6.2) is a statement of the self-adjointness of this operator, and has as one of its important consequences the following relation:

$$G^{-ij} = G^{+ji} \,, \tag{6.8}$$

where $G^+$ is the advanced Green's function of $S_{,ij}$. Another equation satisfied by $G^+$ and $G^-$, which we shall need, is:

$$G^{\pm ij}{}_{,m} = G^{\pm ik}\, G^{\mp jl}\, S_{,klm} \,. \tag{6.9}$$

Equations (6.8) and (6.9) are derived in Appendix. B.

## 7. – The Poisson bracket.

The addition of the term $\varepsilon r$ to the action $S$ produces retarded disturbances not only in the dynamical variables $q^i$ but also in any observable built out of these variables. Thus, the retarded change in an observable $s$ is given by

$$\delta s = s_{,i}\,\delta q^i = \varepsilon s_{,i} G^{-ij} r_{,j} \,. \tag{7.1}$$

It is convenient to introduce the following notation due to PEIERLS [10]:

$$D_r s = \lim_{\varepsilon \to 0} \frac{1}{\varepsilon}\,\delta s = s_{,i} G^{-ij} r_{,j} \,. \tag{7.2}$$

The Poisson bracket of the two observables $r$ and $s$ may then be defined very succinctly as

$$(r, s) = D_r s - D_s r = s_{,i} G^{-ij} r_{,j} - r_{,i} G^{-ij} s_{,j} = r_{,i} \tilde{G}^{ij} s_{,j} \,, \tag{7.3}$$

where

$$\tilde{G}^{ij} = G^{+ij} - G^{-ij} = -\tilde{G}^{ji} \,. \tag{7.4}$$

We have, in particular,

$$(q^i, q^j) = \tilde{G}^{ij} . \tag{7.5}$$

The familiar antisymmetry of the Poisson bracket is built into the definition (7.3). That this definition also satisfies the well-known law

$$(r, f(s_\alpha)) = (r, s_\alpha) \frac{\partial f}{\partial s_\alpha} , \tag{7.6}$$

follows immediately from the fact that functional differentiation satisfies the same chain rule as does ordinary differentiation. Only the Poisson-Jacobi identity requires some effort to demonstrate. The steps are given in Appendix C.

Once the Poisson bracket has been defined, the modern statement of the correspondence principle may be regarded as the adjunction to the operator dynamical equations of the *quantization rule*

$$[\mathbf{r}, \mathbf{s}] = i(\mathbf{r}, \mathbf{s}) , \qquad \hbar = 1 , \tag{7.7}$$

$\mathbf{r}$ and $\mathbf{s}$ being the operators whose classical analogs are $r$ and $s$, and $(\mathbf{r}, \mathbf{s})$ being an operator (assumed determinable by some uniqueness criterion) whose classical analog is the Poisson bracket $(r, s)$.

As an elementary illustration of the above formalism consider the harmonic oscillator in one dimension. There is one dynamical variable $x(t)$, denoting displacement from equilibrium. The action functional has the form:

$$S = \int \tfrac{1}{2} m(\dot{x}^2 - \omega^2 x)\, dt , \tag{7.8}$$

where $\dot{x} = dx/dt$, and the dynamical equation is

$$0 = \frac{\delta s}{\delta x} = - m(\ddot{x} + \omega^2 x) . \tag{7.9}$$

The operator $S_{,ij}$ takes the form

$$\frac{\delta^2 S}{\delta x(t)\, \delta x(t')} = - m \left( \frac{\partial^2}{\partial t^2} + \omega^2 \right) \delta(t - t') \tag{7.10}$$

and possesses the retarded Green's function

$$G^-(t, t') = \frac{1}{m\omega} \theta(t - t') \sin \omega(t - t') , \tag{7.11}$$

where $\theta$ is the « switch-on » step function. From this, Poisson brackets may be computed immediately. For example,

$$(x(t), x(t')) = \tilde{G}(t, t') = -\frac{1}{m\omega}\sin\omega(t-t')\,, \tag{7.12}$$

which yileds, in particular, the equal-time Poisson bracket

$$(x(t), p(t)) = 1\,, \qquad p = m\dot{x}\,. \tag{7.13}$$

The above expressions also yield, in the limit $\omega\to 0$, the corresponding forms for the free particle:

$$G^-(t, t') = \theta(t-t')\frac{t-t'}{m}\,, \tag{7.14}$$

$$(x(t), x(t')) = -\frac{t-t'}{m}\,. \tag{7.15}$$

In its quantum form eq. (7.15) may be interpreted as a statement of the wave-packet-spreading phenomenon.

## 8. – Measurement of a single observable. Uncertainties and compensation devices.

In order to make a measurement on the system $S$ we must introduce an apparatus and an appropriate coupling. In the absence of the coupling the combined action functional for the system and apparatus will have the form $S[q]+\Sigma[Q]$, where $\Sigma$ is the action functional of the apparatus alone and the symbols $Q^I$ denote its dynamical variables. We shall assume we know the approximate initial state of the apparatus and hence, to some degree of accuracy, what the whole trajectory of the apparatus would be if it remained uncoupled to the system. This trajectory will be given by functions $Q^I(t)$ satisfying the dynamical equations

$$\Sigma_{,I}[Q] = 0\,. \tag{8.1}$$

The (at least partially) unknown trajectory of the system is given similarly by (unknown) functions $q^i(t)$ satisfying eq. (6.1).

The introduction of a coupling term into the total action functional produces disturbances in both the system and the apparatus, *i.e.*, deviations from the trajectories $q^i(t)$, $Q^I(t)$. The disturbance in the apparatus is what makes the measurement possible. The disturbance in the system, on the other hand, tends to change the physical quantity under observation and hence to complicate the measurement one is trying to make. For this reason one imagines, in classical physics, that the coupling can be made as weak as desired. However, the weaker the coupling, the harder is it to detect the disturbance in the apparatus. Therefore a detailed study of the measurement process must be carried out before an accurate judgment can be rendered concerning limits of error. It turns out that although these limits are only of a practical nature in the classical theory, in the quantum theory they are fundamental.

Suppose we want to measure the system observable $s$. What coupling term shall we choose for this purpose? In more technical language, what shall we add to the total action $S + \Sigma$ in order to produce disturbances in the apparatus that, in the quantum theory, are described by the unitary transformation (1.5)? As a first guess let us try simply $g\mathscr{X}$ itself, so that the change in the action becomes:

$$S + \Sigma \to S + \Sigma + g\mathscr{X}. \tag{8.2}$$

In the weak-coupling limit the resulting changes in the dynamical trajectories may be obtained by applying the theory of small disturbances introduced in Sect. 6. More generally, we may introduce a functional Taylor expansion. Thus, letting the actual (disturbed) trajectories be denoted by $q^i + \delta q^i$, $Q^I + \delta Q^I$, we have

$$\begin{aligned} 0 &= S_{,i}[q + \delta q] + g\mathscr{X}_{,i}[q + \delta q, Q + \delta Q] = \\ &= S_{,ij}\delta q^j + \tfrac{1}{2}S_{,ijk}\delta q^j \delta q^k + \ldots + g\mathscr{X}_{,i} + g\mathscr{X}_{,ij}\delta q^j + g\mathscr{X}_{,iJ}\delta Q^J + \ldots, \end{aligned} \tag{8.3}$$

$$\begin{aligned} 0 &= \Sigma_{,I}[Q + \delta Q] + g\mathscr{X}_{,I}[q + \delta q, Q + \delta Q] = \\ &= \Sigma_{,IJ}\,\delta Q^J + \tfrac{1}{2}\Sigma_{,IJK}\delta Q^J \delta Q^K + \ldots + g\mathscr{X}_{,I} + g\mathscr{X}_{,Ii}\delta q^i + g\mathscr{X}_{,IJ}\delta Q^J + \ldots, \end{aligned} \tag{8.4}$$

where the coefficients in the expansions are evaluated at the undisturbed trajectories.

Equations (8.3) and (8.4) may be solved by iteration. Assuming that $\delta q^i$ and $\delta Q^I$ satisfy retarded boundary conditions, we obtain from eq. (8.3), to lowest order in $g$,

$$\delta q^i = gG^{-ij}\,\mathscr{X}_{,j}, \tag{8.5}$$

where $G^{-ij}$ is the retarded Green's function introduced in Sect. 6. When $g$ is small eq. (8.5) gives the dominant contribution to the disturbance in the system produced by the coupling. In calculating the disturbance in the apparatus we must allow for this disturbance in the system. Hence, consistency requires us to solve eq. (8.4) correct to *second* order in $g$. However, we can eliminate some of the second-order terms by assuming the apparatus to be much more massive than the system. Equations (7.11) and (7.14) show that Green's functions typically depend inversely on the mass. Hence, under this assumption (which, incidentally, is the only feature of macroscopicality I shall assign to the apparatus) we may neglect iteration terms involving the retarded Green's function $G^{-IJ}$ of $\Sigma_{,IJ}$ in comparison with analogous terms involving $G^{-ij}$. This yields

(8.6) $$\delta Q^I = gG^{-IJ}(\mathscr{X}_{,J} + \mathscr{X}_{,Ji}\delta q^i) = gG^{-IJ}(\mathscr{X}_{,J} + g\mathscr{X}_{,Ji}G^{-ij}\,\mathscr{X}_{,j})\,.$$

Now let $A$ be an arbitrary apparatus observable. The change which the coupling $g\mathscr{X}$ produces in $A$ may be expressed compactly with the aid of the Peierls notation introduced in the preceding Section:

(8.7) $$\delta A = A_{,I}\delta Q^I = g(1 + gD_{\mathscr{X}})D_{\mathscr{X}}A\,,$$

where

(8.8) $$D_{\mathscr{X}}D_{\mathscr{X}}A = (D_{\mathscr{X}}A)_{,i}G^{-ij}\mathscr{X}_{,j} = A_{,I}G^{-IJ}\mathscr{X}_{,Ji}G^{-ij}\mathscr{X}_{,j}\,,$$

the term $(D_{\mathscr{X}}A)_{,I}G^{-IJ}\mathscr{X}_{,J}$ that would normally be included in the definition of $D_{\mathscr{X}}D_{\mathscr{X}}A$ being omitted because of the massiveness of the apparatus. Note that all quantities in the above expressions are to be evaluated using the undisturbed trajectories of system and apparatus.

Suppose $A$ is the apparatus observable introduced in Sect. 1. It is then built out of $Q$'s taken from an interval of time lying to the future of the time interval associated with the dynamical variables out of which $\mathscr{X}$ is constructed. This implies

(8.9) $$D_A\mathscr{X} = 0$$

and hence

(8.10) $$D_{\mathscr{X}}A = (\mathscr{X}, A) = s\,,$$

where the classical analog of eq. (1.6) has been used in the final steps. Equations (8.7) and (8.10) together now yield for the disturbed apparatus observable

(8.11) $$\bar{A} = A + \delta A = A + gs + g^2D_{\mathscr{X}}s\,.$$

$\bar{A}$ is seen to differ from the classical analog of expression (1.8) only by a term in $g^2$. In the weak-coupling limit one might suppose that this term may be neglected and that our comparison of the quantum and classical descriptions of measurement may stop at this point. If this were so we should have learned very little, because our work up to now has amounted to hardly more than an elementary exercise. In fact, however, the term in $g^2$ cannot be neglected. It has an important influence on the accuracy with which $s$ can be determined, as will now be shown.

If we solve eq. (8.11) to obtain the «experimental value» of $s$, we find

$$s = \frac{\delta A}{g} - gD_{\mathfrak{X}}s\,. \tag{8.12}$$

As has been remarked in Sect. 1, the change $\delta A$ in $A$ constitutes a storage, *in the apparatus*, of information concerning the value of $s$. Since the time interval associated with $A$ lies to the future of the coupling interval, this information could in principle be read out of the apparatus (by another apparatus!) at any later time without further affecting the system. That is to say, $A + \delta A$ could be determined with arbitrary accuracy, and the accuracy with which $s$ would be thereby determined would depend only upon the accuracy with which the undisturbed trajectories would have been known, *i.e.* if the coupling had not been present.

Let us assume for the present that $D_{\mathfrak{X}}s$ depends only on apparatus variables. Then the accuracy with which eq. (8.12) determines $s$ depends only upon the accuracy with which the undisturbed *apparatus* trajectory is known. Denote by $\Delta A$ and $\Delta D_{\mathfrak{X}}s$ the uncertainties in our knowledge of the undisturbed $A$ and $D_{\mathfrak{X}}s$ respectively. The mean square error in the experimental value of $s$ that these uncertainties generate is then given by

$$(\Delta s)^2 = \frac{1}{2}\left(\frac{\Delta A}{g} - g\Delta D_{\mathfrak{X}}s\right)^2 + \frac{1}{2}\left(\frac{\Delta A}{g} + g\Delta D_{\mathfrak{X}}s\right)^2 = \frac{(\Delta A)^2}{g^2} + g^2(\Delta D_{\mathfrak{X}}s)^2\,. \tag{8.13}$$

We see at once from this equation how the error $\Delta s$ behaves as the coupling constant $g$ is varied. When $g$ is very large $\Delta s$ is large due to the uncertainty in the disturbance $gD_{\mathfrak{X}}s$ produced in the system. (Note that it is the *uncertainty* in the disturbance which is important here and not the disturbance itself, which could in principle be allowed for.) When $g$ is very small, on the other hand, $\Delta s$ again becomes large because of the difficulty of obtaining a meaningful value for $\delta A$. (It gets swamped by the uncertainty $\Delta A$.) The minimum value of $\Delta s$ occurs for

$$g^2 = \frac{\Delta A}{\Delta D_{\mathfrak{X}}s}\,, \tag{8.14}$$

at which coupling strength we have

$$\Delta s = (2\Delta A\,\Delta D_{\mathfrak{X}}s)^{\frac{1}{2}}\,. \tag{8.15}$$

In the classical theory $\Delta A$ and $\Delta D_{\mathcal{X}} s$ can in principle be made as small as desired. In the quantum theory, however, they are limited by the Heisenberg-Robertson-Schrödinger (HRS) uncertainty relation which, in its quasi-classical form, may be written

$$\Delta A \Delta D_{\mathcal{X}} s \geq \tfrac{1}{2} |(D_{\mathcal{X}} s, A)| , \qquad \hbar = 1 . \tag{8.16}$$

But

$$(D_{\mathcal{X}} s \;\; A) = D_{D_{\mathcal{X}} s} A = A_{,I} G^{-IJ} s_{,i} G^{-ij} \mathcal{X}_{,Ji} = D_{D_{\mathcal{X}} A} s = D_s s \tag{8.17}$$

and therefore we conclude that

$$\Delta s \geq |D_s s|^{\frac{1}{2}} . \tag{8.18}$$

Because all reference to the apparatus has now disappeared, this inequality appears to suggest that there is, in the quantum theory, a fundamental limit to the accuracy with which the value of any *single* observable may be known. This, however, is in direct conflict with the well established principle that the value of any single observable is determinable with arbitrary accuracy even in quantum mechanics.

The way to overcome this apparent contradiction was discovered by BOHR and ROSENFELD [2]. They simply modified the coupling between system and apparatus by inserting an additional term $-\frac{1}{2} g^2 D_{\mathcal{X}} \mathcal{X}$ into the total action so that (8.2) gets replaced by

$$S + \Sigma \rightarrow S + \Sigma + g\mathcal{X} - \tfrac{1}{2} g^2 D_{\mathcal{X}} \mathcal{X} . \tag{8.19}$$

The addition of this term, which is known as a *compensation term*, has the effect of replacing eq. (8.7) by

$$\delta A = g D_{\mathcal{X}} A + g^2 D_{\mathcal{X}} D_{\mathcal{X}} A - \tfrac{1}{2} g^2 D_{D_{\mathcal{X}} \mathcal{X}} A . \tag{8.20}$$

In the approximation of neglecting terms containing the Green's function $G^{-IJ}$ in comparison with analogous terms containing $G^{-ij}$, the last two terms of this equation may be rewritten in the form

$$\begin{aligned} D_{\mathcal{X}} D_{\mathcal{X}} A - \tfrac{1}{2} D_{D_{\mathcal{X}} \mathcal{X}} A &= (D_{\mathcal{X}} A)_{,i} G^{-ij} \mathcal{X}_{,j} - \tfrac{1}{2} A_{,I} G^{-IJ} (D_{\mathcal{X}} \mathcal{X})_{,J} = \\ &= A_{,I} G^{-IJ} (\mathcal{X}_{,Ji} G^{-ij} \mathcal{X}_{,j} - \tfrac{1}{2} \mathcal{X}_{,Ji} G^{-ij} \mathcal{X}_{,j} - \tfrac{1}{2} \mathcal{X}_{,j} G^{-ji} \mathcal{X}_{,Ji}) = \\ &= -\tfrac{1}{2} A_{,I} G^{-IJ} \mathcal{X}_{,Ji} \tilde{G}^{ij} \mathcal{X}_{,j} = -\tfrac{1}{2} (D_{\mathcal{X}} A, \mathcal{X}) = \tfrac{1}{2} (\mathcal{X}, s) , \end{aligned} \tag{8.21}$$

in which use has been made of eqs. (6.8), (7.3), (7.4) and (8.10). We now invoke the classical analog of eq. (1.7), namely,

$$(8.22) \qquad (\mathscr{X}, s) = 0\,,$$

which tells us that the last two terms of eq. (8.20) actually cancel one another, leaving

$$(8.23) \qquad \delta A = g D_{\mathscr{X}} A = g s\,,$$

and hence

$$(8.24) \qquad s = \frac{\delta A}{g}\,, \qquad \Delta s = \frac{\Delta A}{g}\,.$$

With the introduction of a compensation term the uncertainty $\Delta s$ can evidently be made arbitraryly small, either by making $g$ large enough (but not so big that higher order terms in the Taylor expansions (8.3) and (8.4) must be retained) or by taking $\Delta A$ small enough. In the quantum theory the latter condition is expedited by keeping the mass of the apparatus big. If the spectrum of $s$ is discrete, as we assumed in Part I, then of course a completely precise experimental value for $s$ can be obtained merely by requiring $\Delta A$ to satisfy the condition (1.22).

Some comments are now in order regarding the idealizations that have been made thus far. First of all, it should be recognized that no real apparatus is designed to make a successful measurement regardless of the trajectory it finds itself in and regardless of the state of the system. The success of a measurement usually depends not only on the accuracy with which the undisturbed apparatus trajectory is known but also on a careful choice of this trajectory. For example, the time interval associated with the coupling $g\mathscr{X}$ is often not selected by making $\mathscr{X}$ depend explicitly on the time but by making a special choice of trajectory. (See, for example, the Stern-Gerlach experiment described in the next Section.) Furthermore, relations such as (8.10) and (8.22) will generally not hold as identities for all trajectories but only for certain classes of trajectories. In the quantum theory this means that their analogs, eqs. (1.6) and (1.7), will only hold within a certain subspace of the full Hilbert space (in which $|\Psi\rangle$, of course, lies), and the validity of (1.6) and (1.7) may even depend on some of the suppressed labels having certain values. Finally, in the error analysis presented above it is not really necessary to require that $D_{\mathscr{X}} s$ depend only on apparatus variables. It suffices merely to require that it vary slowly as the system trajectory changes and that the uncertainty in its value stemming from our lack of precise knowledge of the system trajectory be negligible compared to that arising from the imprecision in our knowledge of the apparatus trajectory.

## 9. – Two prototypical measurements.

*A) Stern-Gerlach experiment.* – It is convenient to idealize this experiment by ignoring spin precession. Then the «system» becomes dynamically inert ($S$ = constant), and the value of the $z$ component of the spin, which is the observable $s$ in this case, remains constant and unaffected by the coupling ($D_s s = 0$). The possible values of this constant are discrete and finite in number. The «apparatus» is the atom itself (mass $m$) together with an inhomogeneous magnetic field oriented in the $z$-direction, the apparatus which produces the field, and an associated co-ordinate system. The dynamical variables of the «apparatus» are the Cartesian co-ordinates $x$, $y$, $z$ of the atom, while the dynamical variable of the «system» is the inert spin value. Because $S$ is merely a constant the total undisturbed action functional is effectively that of the apparatus alone:

$$\Sigma = \int \tfrac{1}{2} m(\dot{x}^2 + \dot{y}^2 + \dot{z}^2)\,\mathrm{d}t \,. \tag{9.1}$$

In the absence of coupling (*i.e.* when the magnetic field is switched off) the trajectory that the apparatus (*i.e.* the atom) would follow will be assumed to be given approximately by:

$$x = 0\,, \qquad y = vt\,, \qquad z = 0\,. \tag{9.2}$$

That is, the atom moves along the $y$ axis with velocity $v$, passing the origin at time $t = 0$. We need not inquire how the atom was prepared in this state. We only remark that if the atom is sufficiently massive it will not deviate greatly from the trajectory (9.2), either as a result of quantum-mechanical spreading or as a result of coupling with the magnetic field, until it has passed well beyond the region of nonvanishing field. The magnet itself will be assumed to surround the segment $0 < y < L$ of the $y$-axis. Under these conditions the magnetic field will, in the region traversed by the atom, be expressible to good approximation in the form

$$H_x = 0\,, \qquad H_y = 0\,, \qquad H_z = \theta(y)\theta(L-y)(\alpha + \beta z)\,. \tag{9.3}$$

The coupling term therefore becomes effectively

$$g\mathscr{X} = \int \mu s H_z \,\mathrm{d}t = \mu s \int \theta(y)\theta(L-y)(\alpha + \beta z)\,\mathrm{d}t\,, \tag{9.4}$$

where $\mu$ is the magnetic moment of the atom. If we make the identification

$$g = \mu\beta L/v\,, \tag{9.5}$$

then we may express $\mathscr{X}$ in the special form (1.9) with

$$X = \frac{v}{L}\int \theta(y)\theta(L-y)\left(\frac{\alpha}{\beta}+z\right)\mathrm{d}t\,. \tag{9.6}$$

In the present case the apparatus observable $A$ is conveniently taken to be the $z$-component of momentum of the atom after it has passed through the magnetic field:

$$A = [m\dot{z}]_{t=L/v} = [m\dot{z}]_{t>L/v}\,. \tag{9.7}$$

Using the Poisson-bracket relation (7.15), applied to the $z$-component of position of the atom, one easily finds

$$(X, A) = \frac{v}{L}\int \theta(y)\theta(L-y)\mathrm{d}t = \frac{1}{L}\int_0^L \mathrm{d}y = 1\,, \tag{9.8}$$

in which the trajectory eqs. (9.2) are used in the final steps. Equation (9.8) is just the classical analog of eq. (1.10), and hence we see that the Stern-Gerlach experiment constitutes a measurement of precisely the von Neumann type. Because $A = 0$ for the undisturbed trajectory (9.2) we may write the « experimental value » of the spin in the form

$$s = \frac{\bar{A}}{g} = \left[\frac{mv\dot{\bar{z}}}{\mu\beta L}\right]_{t>L/v}, \tag{9.9}$$

where $\bar{z}$ refers to the disturbed trajectory. We note finally that, from the point of view of the simplified notation employed in Sect. 1, the $x$ and $y$ co-ordinates of the atom are suppressed labels, and that this is a case in which the validity of eq. (1.6) depends very much on these labels having the values specified by the trajectory (9.2)

*B) Electric-field measurement* [2]. In this case the system is the whole electromagnetic field, and its action functional is

$$S = -\tfrac{1}{4}\int F_{\mu\nu}F^{\mu\nu}\mathrm{d}^4x\,, \tag{9.10}$$

where

$$F_{\mu\nu} = \frac{\partial \mathscr{A}_\nu}{\partial x^\mu} - \frac{\partial \mathscr{A}_\mu}{\partial x^\nu}\,, \qquad (\mathscr{A}_\mu) = (-\varphi, \mathscr{A}) \tag{9.11}$$

$$\mathrm{d}^4x = \mathrm{d}x_0\mathrm{d}x^1\mathrm{d}x^2\mathrm{d}x^3 = \mathrm{d}t\mathrm{d}^3r\,, \quad t = x_0\,, \quad r = (x^1, x^2, x^3)\,, \tag{9.12}$$

and where Greek indices are raised and lowered by means of the Minkowski metric $(\eta^{\mu\nu}) = (\eta_{\mu\nu}) = \text{diag}(-1, 1, 1, 1)$. The apparatus will be chosen to consist of two interpenetrable rigid bodies, initially and finally occupying the same volume $V$, one of which is permanently fixed (relative to the co-ordinates $x^1$, $x^2$, $x^3$) and carries a uniform charge density $\varrho$ while the other is movable during the time interval $T$ (only) and carries a uniform charge density $\varrho$. If the mass $M$ of the movable body is big enough so that its dynamics may be described nonrelativistically, then the apparatus action may be taken in the form

$$\Sigma = \tfrac{1}{2} M \int_T \dot{\boldsymbol{R}}^2 \mathrm{d}t + \text{rotational action}, \tag{9.13}$$

where $\boldsymbol{R}$ is the center of mass of the body. We do not write out the rotational part of the action explicitly since the rotational variables will be suppressed labels in this case.

It will be noted that switching devices (to hold and release the movable body) are needed here, whereas they were not needed in the Stern-Gerlach case. The action therefore has an explicit time dependence (through the presence of $T$). Time independence could in principle be restored by including the switching devices in the total action. The choice of time interval would then be governed by initial conditions on the apparatus motion, just as in the Stern-Gerlach case.

The coupling between the electromagnetic field and the apparatus is

$$g\mathscr{X} = \int j^\mu \mathscr{A}_\mu \mathrm{d}^4 x\,, \tag{9.14}$$

where $j^\mu$ is the charge-current density 4-vector of the apparatus. The undisturbed trajectory of the apparatus will be assumed to be given approximately by

$$\boldsymbol{R} = 0\,. \tag{9.15}$$

If the movable body is sufficiently massive it will not deviate greatly from this trajectory either as a result of quantum-mechanical spreading or as a result of the presence of an electromagnetic field. Under these conditions the 4-vector $(j^\mu) = (j^0, \boldsymbol{j})$ may be approximated by

$$j^0(t, \boldsymbol{r}) = \varrho[\chi_V(\boldsymbol{r}-\boldsymbol{R}) - \chi_V(\boldsymbol{r})]\chi_T(t) = -\varrho\boldsymbol{R}\cdot\nabla\chi_V(\boldsymbol{r})\chi_T(t)\,, \tag{9.16}$$

$$\boldsymbol{j}(t, \boldsymbol{r}) = \varrho\dot{\boldsymbol{R}}\chi_V(\boldsymbol{r})\chi_T(t) + \varrho\boldsymbol{R}\chi_V(\boldsymbol{r})\dot{\chi}_T(t)\,, \tag{9.17}$$

where $\chi$ is the «characteristic function» of set theory:

$$\chi_V(\boldsymbol{r}) = \begin{cases} 1, & \boldsymbol{r}\in V \\ 0, & \boldsymbol{r}\notin V, \end{cases} \qquad \chi_T(t) = \begin{cases} 1, & t\in T \\ 0, & t\notin T. \end{cases} \tag{9.18}$$

The second term on the right of eq. (9.17) arises from the possibility of sudden motions of the body caused by the switching devices that act at the beginning and end of the time interval $T$.

Equations (9.16) and (9.17) can be rewritten in the form

$$j^0 = -\nabla\cdot\boldsymbol{P}, \qquad \boldsymbol{j} = \dot{\boldsymbol{P}},$$

where

(9.19) $$\boldsymbol{P} = \varrho\boldsymbol{R}\chi_V(\boldsymbol{r})\,\chi_T(t)\,.$$

More generally, in an arbitrary co-ordinate frame, we may write

(9.20) $$j^\mu = \partial P^{\mu\nu}/\partial x^\nu\,,$$

where $P^{\mu\nu}$ is an antisymmetric polarization tensor. In the co-ordinate system $(t, \boldsymbol{r})$ this tensor takes the form

(9.21) $$(P^{\mu\nu}) = \begin{pmatrix} 0 & -\boldsymbol{P} \\ \boldsymbol{P} & 0 \end{pmatrix}.$$

It will be noted that (9.20) guarantees charge conservation:

(9.22) $$\partial j^\mu/\partial x^\mu = 0\,.$$

If we insert (9.20) into (9.14) and integrate by parts, we get

(9.23a) $$g\,\mathscr{X} = \tfrac{1}{2}\int P^{\mu\nu}F_{\mu\nu}\,\mathrm{d}^4x\,,$$

(9.23b) $$= \int \boldsymbol{P}\cdot\boldsymbol{E}\,\mathrm{d}^4x\,,$$

where $\boldsymbol{E}$ is the electric field vector:

(9.24) $$\boldsymbol{E} = -\nabla\varphi - \dot{\mathscr{A}}\,.$$

Inserting (9.19) into (9.23b) and making the identification

(9.25) $$g = \varrho VT\,,$$

we then find

(9.26) $$\mathscr{X} = \frac{1}{T}\int_T \boldsymbol{R}\cdot\boldsymbol{E}_V\,\mathrm{d}t\,,$$

where $E_V$ is the average of the electric vector over the volume $V$:

$$E_V = \frac{1}{V}\int_V E\,d^3r\,. \tag{9.27}$$

For the apparatus observable $A$ we choose in this case

$$A = [M\dot{R}]_{t=\tau}\,, \tag{9.28}$$

where $\tau$ is the time at the end of the inverval $T$, just before the switching device brings the movable body to rest again. The quantity that the apparatus then measures is a space-time average of the electric field:

$$E_{VT} = \frac{1}{T}\int_T E_V\,dt = \frac{1}{VT}\int_T dt\int_V d^3r\,E\,, \tag{9.29}$$

for it is easy to verify that

$$(\mathscr{X}, A) = E_{VT}\,. \tag{9.30}$$

To get an accurate measurement of the indisturbed $E_{VT}$, however, it is necessary in the present case to supplement the apparatus with a compensation device. This is because any uncertainty that may exist about the apparatus trajectory's being given *precisely* by eq. (9.15) is propagated to the electromagnetic field through the fact that we cannot be sure that $\boldsymbol{P}$, and hence the field produced by the apparatus itself, is exactly zero. To determine the structure of the compensation device required we must first study the dynamics of the electromagnetic field.

The undisturbed field $F_{\mu\nu}$ satisfies the differential equation

$$\partial F^{\mu\nu}/\partial x^\nu = 0\,, \tag{9.31}$$

while the disturbed field $\bar{F}_{\mu\nu}$ satisfies the equation

$$\partial \bar{F}^{\mu\nu}/\partial x^\nu = j^\mu\,. \tag{9.32}$$

The difference between these two equations yields

$$\partial\, \delta F^{\mu\nu}/\partial x^\nu = j^\mu\,, \tag{9.33}$$

which, when combined with the equation

$$\partial\, \delta F_{\mu\nu}/\partial x^\sigma + \partial\, \delta F_{\nu\sigma}/\partial x^\mu + \partial\, \delta F_{\sigma\mu}/\partial x^\nu = 0 \tag{9.34}$$

(which expresses the fact that the field tensor is a curl), yields

$$\Box\delta F_{\mu\nu} = \partial j_\mu/\partial x^\nu - \partial j_\nu/\partial x^\mu, \tag{9.35a}$$

$$= \partial^2 P_\mu{}^\sigma/\partial x^\sigma \partial x^\nu - \partial^2 P_\nu{}^\sigma/\partial x^\sigma\partial x^\mu. \tag{9.35b}$$

The retarted solution of this equation is

$$\delta F_{\mu\nu} = -\int G^-(x-x')(\partial^2 P'^{\sigma}_{\mu}/\partial x'^\sigma \partial x'^\nu - \partial^2 P'^{\sigma}_{\nu}/\partial x'^\sigma\partial x'^\mu)\,\mathrm{d}^4x' = \tag{9.36}$$

$$= -\frac{1}{2}\int P'^{\sigma\tau}\left(\eta_{\mu\sigma}\frac{\partial^2}{\partial x^\nu\partial x^\tau} - \eta_{\mu\tau}\frac{\partial^2}{\partial x^\nu\partial x^\sigma} - \eta_{\nu\sigma}\frac{\partial^2}{\partial x^\mu\partial x^\tau} + \eta_{\nu\tau}\frac{\partial^2}{\partial x^\mu\partial x^\sigma}\right) G^-(x-x')\,\mathrm{d}^4x',$$

where $G^-$ is the retarded Green's function of the d'Alembertian operator $\Box$:

$$G^-(x) = \frac{1}{2\pi}\theta(x^0)\,\delta(x_\mu x^\mu) = \frac{1}{4\pi r}\delta(t-r), \qquad r = |\boldsymbol{r}|. \tag{9.37}$$

On the other hand, we have

$$\delta F_{\mu\nu} = D_{g\mathscr{X}} F_{\mu\nu} = \tfrac{1}{2}\int P'^{\sigma\tau} D_{F'_{\sigma\tau}} F_{\mu\nu}\,\mathrm{d}^4x', \tag{9.38}$$

the second form following from eq. (9.23*a*) and the superposability of electromagnetic-field disturbances. Comparison of eqs. (9.36) and (9.38) therefore yields

$$D_{F'_{\sigma\tau}} F_{\mu\nu} = \tag{9.39}$$

$$= -\left(\eta_{\mu\sigma}\frac{\partial^2}{\partial x^\nu\partial x^\tau} - \eta_{\mu\tau}\frac{\partial^2}{\partial x^\nu\partial x^\sigma} - \eta_{\nu\sigma}\frac{\partial^2}{\partial x^\mu\partial x^\tau} + \eta_{\nu\tau}\frac{\partial^2}{\partial x^\mu\partial x^\sigma}\right) G^-(x-x'),$$

and, in particular,

$$D_{E'}E = \left(\nabla\nabla - \mathbf{1}\frac{\partial^2}{\partial t^2}\right) G^-(x-x'), \tag{9.40}$$

where $\mathbf{1}$ is the unit dyadic.

It is now a straightforward matter to compute the compensation term that needs to be added to the total action functional. Ignoring the Green's function of the apparatus compared to that of the system, we have

$$\left|\begin{aligned} -\tfrac{1}{2}g^2 D_{\mathscr{X}}\mathscr{X} &= -\tfrac{1}{2}\int \mathrm{d}^4x\int \mathrm{d}^4x'\,\boldsymbol{P}\cdot D_{\boldsymbol{E}}\boldsymbol{E}'\cdot\boldsymbol{P}', \\ &= -\tfrac{1}{2}\varrho^2\int_{VT}\mathrm{d}^4x\int_{VT}\mathrm{d}^4x'\,\boldsymbol{R}\cdot D_{\boldsymbol{E}}\boldsymbol{E}'\cdot\boldsymbol{R}'. \end{aligned}\right. \tag{9.41}$$

If the movable body is sufficiently massive $\boldsymbol{R}$ will not change appreciably during the time interval $T$, and (9.41) will be adequately approximated by

$$-\tfrac{1}{2}g^2 D_{\mathscr{X}}\,\mathscr{X} = -\tfrac{1}{2}\int_T \boldsymbol{R}\cdot\boldsymbol{\varkappa}\cdot\boldsymbol{R}\,\mathrm{d}t\,, \tag{9.42}$$

where

$$\boldsymbol{\varkappa} = \frac{\varrho^2}{T}\int_{VT}\mathrm{d}^4x\int_{VT}\mathrm{d}^4x'\left(\nabla\nabla - 1\,\frac{\partial^2}{\partial t^2}\right)G^-(x-x')\,. \tag{9.43}$$

Expression (9.42) is just the contribution to the action functional that would be made by the potential energy term of the Lagrangian of a 3-dimensional harmonic oscillator having spring constant $\boldsymbol{\varkappa}$. This is the origin of the famous mechanical springs that BOHR and ROSENFELD [2] found it necessary to attach to the movable body.

## 10. – Measurement of two observables.

The relationship between the classical and quantum descriptions of the measurement process stands fully revealed only when we examine the complications that arise when the apparatus tries to measure two system observables, $r$ and $s$. In view of what we have learned in earlier Sections we naturally attempt to accomplish such a measurement via a coupling of the form

$$\begin{aligned} &g(\mathscr{X}+\mathscr{Y}) - \tfrac{1}{2}g^2 D_{(\mathscr{X}+\mathscr{Y})}(\mathscr{X}+\mathscr{Y}) = \\ &\qquad = g(\mathscr{X}+\mathscr{Y}) - \tfrac{1}{2}g^2(D_{\mathscr{X}}\,\mathscr{X} + D_{\mathscr{X}}\,\mathscr{Y} + D_{\mathscr{Y}}\,\mathscr{X} + D_{\mathscr{Y}}\,\mathscr{Y})\,, \end{aligned} \tag{10.1}$$

in which a Bohr-Rosenfeld compensation term has been included. Here $\mathscr{X}$ and $\mathscr{Y}$ are required to satisfy

$$D_{\mathscr{X}}A = (\mathscr{X}, A) = r\,, \qquad (\mathscr{X}, r) = 0\,, \tag{10.2}$$

$$D_{\mathscr{Y}}B = (\mathscr{Y}, B) = s\,, \qquad (\mathscr{Y}, s) = 0\,, \tag{10.3}$$

$$D_{\mathscr{X}}B = (\mathscr{X}, B) = 0\,, \qquad D_{\mathscr{Y}}A = (\mathscr{Y}, A) = 0\,, \tag{10.4}$$

where $A$ and $B$ are the apparatus observables that store the observations of $r$ and $s$ respectively, and the Poisson brackets are to be evaluated at the undisturbed trajectories. It will be noted that the inclusion of an overall compensation term for $\mathscr{X}+\mathscr{Y}$ is equivalent to the inclusion of individual compensation terms for $\mathscr{X}$ and $\mathscr{Y}$ separately, plus another term, $-\frac{1}{2}g^2(D_{\mathscr{X}}\mathscr{Y}+D_{\mathscr{Y}}\mathscr{X})$, which may be called a *correlation term.*

In addition to eqs. (10.2), (10.3) and (10.4), $A$ and $B$ themselves will be required to satisfy

$$D_A B = D_B A = 0 . \tag{10.5}$$

Equations (10.4) and (10.5) together express the fact that $A$ and $B$ are to be dynamically independent quantities and that the two measurements are to be operations which are as independent of one another as possible, so that any interference that occurs between the measurements of $r$ and $s$ arises not from the memory storage process (*i.e.* from the apparatus) but from the dynamical properties of the system alone. It should be noted that we cannot require the Poisson brackets $(\mathscr{X}, s)$ and $(\mathscr{Y}, r)$ to vanish unless $(r, s) = 0$, for this would contradict the relations

$$(A, (\mathscr{X}, s)) = -(\mathscr{X}, (s, A)) - (s, (A, \mathscr{X})) = (s, r) , \tag{10.6}$$

$$(B, (\mathscr{Y}, r)) = -(\mathscr{Y}, (r, B)) - (r, (B, \mathscr{Y})) = (r, s) . \tag{10.7}$$

We can, however, require

$$((\mathscr{X}, s), (\mathscr{Y}, r)) = 0 , \tag{10.8}$$

as may be seen by making the special choices (cf. eqs. (1.9) and (1.10))

$$\mathscr{X} = rX , \qquad \mathscr{Y} = sY , \tag{10.9}$$

where $X$ and $Y$ are apparatus observables satisfying

$$(X, A) = 1 , \qquad (Y, B) = 1 , \tag{10.10}$$

$$D_X Y = D_Y X = 0 . \tag{10.11}$$

By a series of steps completely analogous to those followed in obtaining eq. (8.21), it is easy to verify that the coupling (10.1) produces the following disturbances in $A$ and $B$:

$$\delta A = g(\mathscr{X} + \mathscr{Y}, A) + \tfrac{1}{2}g^2(\mathscr{X} + \mathscr{Y}, (\mathscr{X} + \mathscr{Y}, A)) = gr + \tfrac{1}{2}g^2(\mathscr{Y}, r) , \tag{10.12}$$

$$\delta B = g(\mathscr{X} + \mathscr{Y}, B) + \tfrac{1}{2}g^2(\mathscr{X} + \mathscr{Y}, (\mathscr{X} + \mathscr{Y}, B)) = gs + \tfrac{1}{2}g^2(\mathscr{X}, s) , \tag{10.13}$$

correct to second order in $g$. Solving these equations for the «experimental values» of $r$ and $s$, we find:

$$r = \frac{\delta A}{g} - \frac{1}{2} g(\mathscr{Y}, r)\,, \tag{10.14}$$

$$s = \frac{\delta B}{g} - \frac{1}{2} g(\mathscr{X}, s)\,, \tag{10.15}$$

which lead to the uncertainty relations

$$(\Delta r)^2 = \frac{(\Delta A)^2}{g^2} + \frac{1}{4} g^2 [\Delta(\mathscr{Y}, r)]^2\,, \tag{10.16}$$

$$(\Delta s)^2 = \frac{(\Delta B)^2}{g^2} + \frac{1}{4} g^2 [\Delta(\mathscr{X}, s)]^2\,. \tag{10.17}$$

If we multiply these two equations together we find that the minimum value of the product of the uncertainties $\Delta r$ and $\Delta s$ occurs for a value of the coupling constant given by

$$g^4 = \frac{4 \Delta A \Delta B}{\Delta(\mathscr{Y}, r) \Delta(\mathscr{X}, s)}\,. \tag{10.18}$$

At this value we have:

$$\Delta r \Delta s = \tfrac{1}{2} [\Delta A \Delta(\mathscr{X}, s) + \Delta B \Delta(\mathscr{Y}, r)]\,. \tag{10.19}$$

Under the best imaginable circumstances the Poisson brackets $(\mathscr{X}, s)$ and $(\mathscr{Y}, r)$ will depend at most weakly on the system trajectory, so that the uncertainties $\Delta(\mathscr{X}, s)$ and $\Delta(\mathscr{Y}, r)$ will arise primarily from the imprecision in our knowledge of the apparatus trajectory. Applying the quasi-classical form of the HRS uncertainty relation to these uncertainties and making use of eqs. (10.6) and (10.7), we find

$$\left.\begin{matrix} \Delta A \Delta(\mathscr{X}, s) \\ \Delta B \Delta(\mathscr{Y}, r) \end{matrix}\right\} > \tfrac{1}{2} |(r, s)| \tag{10.20}$$

and hence

$$\Delta r \Delta s > \tfrac{1}{2} |(r, s)|\,. \tag{10.21}$$

That is, we get the HRS uncertainty relation back again at a new level (*)!

This result constitutes a proof of the fundamental consistency of the quantum-mechanical formalism with the theory of measurement. If the laws of quantum mechanics, as embodied in the uncertainty relation, hold for the apparatus, then they are immediately propagated to every system with which the apparatus interacts. The precision with which we can make mutually interfering measurements at any level, is as good as, but no better than, that allowed by the quantum-mechanical uncertainty relations.

## 11. – Imperfect measurements.

Many interactions are constantly taking place in the universe that lead to decompositions of the state vector of the form (2.1) but for which the orthogonality relations (2.3) do not hold (**). Failure of the orthogonality relations, in fact, always occurs to some extent whenever the spectrum of $s$ is continuous. Such interactions do not split the universe cleanly into noninterfering branches. This means that the branching-universe picture given in Sect. **2**, **3** and **4** is an idealization. In order to arive at a full undertanding of the implications of the many-universes interpretation of quantum mechanics one must also consider imperfect measurements.

After an imperfect measurement it is not possible to regard the system as having been prepared in a definite state which may be studied independently of the rest of the universe. The apparatus has not succeeded in bringing about a state in which the vector $|\Psi\rangle$ may, for practical purposes, be regarded as collapsed. In other words, a true split has not occurred.

The split can, however, be *forced*, by bringing in a second apparatus to measure very accurately the disturbed apparatus observable $\bar{A}$. By this device

(*) Bohr and Rosenfeld [2] showed, in their analysis of the simultaneous measurement of two electromagnetic field averages, that if the correlation term $-\frac{1}{2}g^2(D_{\mathscr{X}}\mathscr{Y} + D_{\mathscr{Y}}\mathscr{X})$ is omitted from the coupling, then the inequality (10.21) is changed to the weaker inequality $\Delta r\,\Delta s > \frac{1}{2}(|D_r s| + |D_s r|)$. Equation (10.21) represents the best that one can do. The actual correlation devices that Bohr and Rosenfeld were compelled to introduce to obtain (10.21) consisted of mechanical springs linking the two movable bodies comprising the apparatus.

(**) The student should perhaps be reminded again at this point that reality is not described by the state vector alone, but by the state vector *plus* a set of dynamical operator variables satisfying definite dynamical equations. Decompositions of the form (2.1) are not to be regarded as meaningful if they are merely abstract mathematical exercises in Hilbert-space. Indeed such mathematical decompositions can be performed in an infinity of ways. Only those decompositions are meaningful which reflect the behavior of a concrete dynamical system.

it is possible to give a meaning and an answer to the question: In what sort of state does an imperfect measurement leave the system?

Let us consider the case in which the spectrum of $s$ is continuous. Expression (2.1), which reflects the situation after the measurement, is then modified to read

$$|\Psi\rangle = \int c_s |s\rangle |\Phi[s]\rangle \mathrm{d}s\,, \tag{11.1}$$

in which the summation over $s$ is replaced by an integral. If now a second apparatus is introduced, which makes a very accurate observation of $\bar{A}$, a true splitting will occur. In the branch that corresponds to the eigenvalue $\bar{A}$ the vector $|\Psi\rangle$ will have been effectively collapsed to $\int |s\rangle|\bar{A}\rangle\langle s, \bar{A}|\Psi\rangle \mathrm{d}s$.

The factor $\int |s\rangle\langle s, \bar{A}|\Psi\rangle \mathrm{d}s$ appearing in this collapsed vector may be regarded as the vector corresponding to the state (appropriate to the branch in question) in which the first measurement has left the system. Denoting this vector by $|\psi\rangle_{\bar{A}}$, we find, with the aid of eq. (2.2),

$$|\psi\rangle_{\bar{A}} = \int c_s \Phi(\bar{A} - gs)|s\rangle \mathrm{d}s\,. \tag{11.2}$$

Unless $\Phi$ is an infinitely narrow function (in which case the first measurement would have been perfect) this is not an eigenvector of $s$. It consists, instead, of a superposition of eigenvectors, corresponding to a range of eigenvalues having a spread $\Delta A/g$ and roughly centered around the value $(\bar{A} - \langle A\rangle)/g$ (*). We see therefore that the uncertainty

$$\Delta s = \frac{\Delta A}{g}\,, \tag{11.3}$$

found in Sect. 8 (see (8.24)), with which the first apparatus measures the observable $s$, is reflected in the state in which the system finds itself after the measurement. In the present case the uncertainty in the measured value of the system observable arises from the uncertainty, in the state $|\Phi\rangle$, of the value of the undisturbed apparatus observable $A$, and can be made as small as desired simply by choosing $\Delta A$ sufficiently small. There is an important case, however, in which uncertainties remain no matter how precisely the relevant apparatus observables are known namely, the case considered in Sect. 10 in which the apparatus tries to measure two noncommuting observables. In this case the two measurements unavoidably interfere with one another.

---

(*) If $c_s$ is itself a narrow function, with $\Delta s \ll \Delta A/g$ (*e.g.*, if $|\psi\rangle$ is an eigenvector of $s$), then $|\psi\rangle_{\bar{A}}$ is just a multiple of $|\psi\rangle$, and the first measurement leaves the system state vector unaffected.

For simplicity let us consider the special case in which the commutator $[r, s]$ (or, equivalently, the Poisson bracket $(r, s)$) is a multiple of the identity operator, and $\mathscr{X}$ and $\mathscr{Y}$ have the forms (10.9), with (10.10) and (10.11) holding. The coupling (10.1) then transforms the undisturbed apparatus observables $A$ and $B$ into the disturbed observables:

$$\bar{A} = \exp[-ig(\mathscr{X}+\mathscr{Y})]A\exp[ig(\mathscr{X}+\mathscr{Y})] = A + gr - \tfrac{1}{2}g^2(r, s)Y, \tag{11.4}$$

$$\bar{B} = \exp[-ig(\mathscr{X}+\mathscr{Y})]B\exp[ig(\mathscr{X}+\mathscr{Y})] = B + gs + \tfrac{1}{2}g^2(r, s)X. \tag{11.5}$$

It will be convenient to introduce also the following operators:

$$\hat{r} = \exp[-ig(\mathscr{X}+\mathscr{Y})]r\exp[ig(\mathscr{X}+\mathscr{Y})] = r - g(r, s)Y, \tag{11.6}$$

$$\hat{s} = \exp[-ig(\mathscr{X}+\mathscr{Y})]s\exp[ig(\mathscr{X}+\mathscr{Y})] = s + g(r, s)X. \tag{11.7}$$

(These are *not* equal to the disturbed operators $\bar{r}$ and $\bar{s}$ respectively.)

Now note that one can obtain eight distinct commuting triplets of operators by choosing one from each of the pairs $\{r, s\}$, $\{A, X\}$, $\{B, Y\}$. Any one of the eight possible Hilbert-space bases determined by these triplets can be used in the description of the undisturbed state of the system-plus-apparatus. Because

$$\left\{\begin{aligned} &\exp[-ig(\mathscr{X}+\mathscr{Y})]X\exp[ig(\mathscr{X}+\mathscr{Y})] = X, \\ &\exp[-ig(\mathscr{X}+\mathscr{Y})]Y\exp[ig(\mathscr{X}+\mathscr{Y})] = Y, \end{aligned}\right. \tag{11.8}$$

it follows that eight distinct commuting triplets of operators can also be obtained by choosing one from each of the pairs $\{\hat{r}, \hat{s}\}$, $\{\bar{A}, X\}$, $\{\bar{B}, Y\}$. Any one of the eight possible Hilbert-space bases determined by these triplets can be used in the description of the ***disturbed*** state of the sistem plus apparatus. Because the pairs $\{A, X\}$, $\{B, Y\}$, $\{\bar{A}, X\}$, $\{\bar{B}, Y\}$ are conjugate pairs, the transformation coefficients between the various bases may be taken in the forms

$$\langle X|A\rangle = \langle X|\bar{A}\rangle = (2\pi)^{-\frac{1}{2}}\exp[iXA] = (2\pi)^{-\frac{1}{2}}\exp[iX\bar{A}], \tag{11.9}$$

$$\langle Y|B\rangle = \langle Y|\bar{B}\rangle = (2\pi)^{-\frac{1}{2}}\exp[iYB] = (2\pi)^{-\frac{1}{2}}\exp[iY\bar{B}]. \tag{11.10}$$

Let us assume an uncorrelated initial state for the system and apparatus, so that the total state vector takes the form

$$|\Psi\rangle = |\psi\rangle|\Phi\rangle, \tag{11.11}$$

the Cartesian product decomposition being into the Hilbert subspaces defined by the undisturbed pairs $\{r, s\}$, $\{A, X\}$, $\{B, Y\}$. Let us further assume, for simplicity, that the two memory cells of the apparatus are uncorrelated, so that $\langle A, B|\Phi\rangle$ factors into a product of the form

$$\langle A, B|\Phi\rangle = \Phi_1(A)\Phi_2(B)\,. \tag{11.12}$$

Then we have also

$$\langle A, Y|\Phi\rangle = \Phi_1(A)\tilde{\Phi}_2(Y)\,, \qquad \langle X, B|\Phi\rangle = \tilde{\Phi}_1(X)\Phi_2(B)\,, \qquad \text{etc.} \tag{11.13}$$

where the tilde denotes the Fourier transform:

$$\tilde{\Phi}_1(X) = (2\pi)^{-\frac{1}{2}}\int \exp[iXA]\,\Phi_1(A)\,\mathrm{d}A\,, \tag{11.14}$$

$$\tilde{\Phi}_2(Y) = (2\pi)^{-\frac{1}{2}}\int \exp[iYB]\,\Phi_2(B)\,\mathrm{d}B\,. \tag{11.15}$$

Consider now the basis vectors $|r, A, Y\rangle$. From eqs. (11.4) and (11.6) we find

$$\begin{cases} \bar{A}|r, A, Y\rangle = [A + gr - \frac{1}{2}g^2(r, s)\,Y]\,|r, A, Y\rangle = \\ \qquad\qquad = [\bar{A} + g\hat{r} - \frac{1}{2}g^2(r, s)\,Y]\,|r, A, Y\rangle\,, \end{cases} \tag{11.16}$$

$$\begin{cases} \hat{r}|r, A, Y\rangle = [r - g(r, s)\,Y]\,|r, A, Y\rangle = \\ \qquad\qquad = [\hat{r} - g(r, s)\,Y]\,|r, A, Y\rangle\,, \end{cases} \tag{11.17}$$

the eigenvalue equalities $r = \hat{r}$, $A = \bar{A}$ being used in passing to the final forms.

From this it follows that, apart form an arbitrary phase factor which will be taken equal to unity, we have

$$|r, A, Y\rangle = |\hat{r} - g(r, s)\,Y, \bar{A} + g\hat{r} - \tfrac{1}{2}g^2(r, s)\,Y, Y\rangle\,, \tag{11.18}$$

or, equivalently,

$$|\hat{r}, \bar{A}, Y\rangle = |r + g(r, s)\,Y, A - gr - \tfrac{1}{2}g^2(r, s)\,Y, Y\rangle\,. \tag{11.19}$$

In a similar manner we can infer

$$|s, X, B\rangle = |\hat{s} + g(r, s)X, X, \bar{B} + g\hat{s} + \tfrac{1}{2}g^2(r, s)X\rangle\,, \tag{11.20}$$

$$|\hat{s}, X, \bar{B}\rangle = |s - g(r, s)X, X, B - gs + \tfrac{1}{2}g^2(r, s)X\rangle\,, \tag{11.21}$$

with $s = \hat{s}$, $B = \bar{B}$.

Using eqs. (11.11), (11.13) and (11.19), it is a straightforward matter to decompose the total state vector in terms of the basis vectors $|\hat{r}, \bar{A}, Y\rangle$:

$$|\Psi\rangle = \int d\hat{r} \int d\bar{A} \int dY |\hat{r}, \bar{A}, Y\rangle c_{\hat{r}+g(r,s)Y} \Phi_1(\bar{A} - g\hat{r} - \tfrac{1}{2}g^2(r, s)Y)\tilde{\Phi}_2(Y) , \tag{11.22}$$

where

$$c_r = \langle r|\psi\rangle . \tag{11.23}$$

The decomposition of $|\Psi\rangle$ in terms of the basis vectors $|\hat{s}, X, \bar{B}\rangle$ has a similar structure:

$$|\Psi\rangle = \int d\hat{s} \int dX \int d\bar{B} |\hat{s}, X, \bar{B}\rangle d_{\hat{s}-g(r,s)X} \tilde{\Phi}_1(X) \Phi_2(\bar{B} - g\hat{s} + \tfrac{1}{2}g^2(r, s)X) , \tag{11.24}$$

where

$$d_s = \langle s|\psi\rangle . \tag{11.25}$$

If we now introduce a second apparatus which makes accurate measurements of $\bar{A}$ and $\bar{B}$, thereby forcing the universe to split, we obtain for the effective state vector of the system, in the branch in which the second apparatus has recorded the values $\bar{A}$ and $\bar{B}$ respectively,

$$\begin{aligned} |\psi\rangle_{\bar{A},\bar{B}} &= \int |\hat{r}\rangle\langle\hat{r}, \bar{A}, \bar{B}|\Psi\rangle d\hat{r} \\ &= (2\pi)^{-\frac{1}{2}} \int d\hat{r} \int dY |\hat{r}\rangle \exp[-i\bar{B}Y] c_{\hat{r}+g(r,s)Y} \cdot \\ &\qquad \cdot \Phi_1(\bar{A} - g\hat{r} - \tfrac{1}{2}g^2(r, s)Y) \tilde{\Phi}_2(Y) , \end{aligned} \tag{11.26a}$$

$$\begin{aligned} &= (2\pi)^{-\frac{1}{2}} \int d\hat{s} \int dX |\hat{s}\rangle \exp[-i\bar{A}X] d_{\hat{s}-g(r,s)X} \cdot \\ &\qquad \cdot \tilde{\Phi}_1(x) \Phi_2(\bar{B} - g\hat{s} + \tfrac{1}{2}g^2(r, s)X) . \end{aligned} \tag{11.26b}$$

This vector does not represent either the observable $\hat{r}$ or the observable $\hat{s}$ as having a definite value. The mean square deviations of these observables from their mean values may easily be estimated in the case in which $c_r$ and $d_s$ are slowly varying functions. Denoting by $\Delta A, \Delta B, \Delta X, \Delta Y$ the root mean square deviations defined by the functions $\Phi_1, \Phi_2, \tilde{\Phi}_1, \tilde{\Phi}_2$ respectively, we obtain

$$(\Delta\hat{r})^2 = \frac{(\Delta A)^2}{g^2} + \frac{1}{4} g^2 (r, s)^2 (\Delta Y)^2 , \tag{11.27}$$

$$(\Delta\hat{s})^2 = \frac{(\Delta B)^2}{g^2} + \frac{1}{4} g^2 (r, s)^2 (\Delta X)^2 , \tag{11.28}$$

the first terms on the right of these equations coming from the factors $\Phi_1$ and $\Phi_2$ in the integrands of eqs. (11.26*a*) and (11.26*b*) respectively, and the second terms coming from the factors $\bar{\Phi}_2$ and $\bar{\Phi}_1$.

Now the deviations $\Delta X$ and $\Delta Y$ are limited by the constraints

$$\Delta X \Delta A > \tfrac{1}{2}, \qquad \Delta Y \Delta B > \tfrac{1}{2}. \tag{11.29}$$

Therefore we have

$$(\Delta \hat{r})^2 > \frac{(\Delta A)^2}{g^2} + \frac{1}{16} g^2 \frac{(r, s)^2}{(\Delta B)^2}, \tag{11.30}$$

$$(\Delta \hat{s})^2 > \frac{(\Delta B)^2}{g^2} + \frac{1}{16} g^2 \frac{(r, s)^2}{(\Delta A)^2}. \tag{11.31}$$

The minimum value of the product of the right-hand sides of these inequalities occurs for a value of the coupling constant given by

$$g^2 = 4 \frac{\Delta A \Delta B}{|(r, s)|}, \tag{11.32}$$

and from this one may readily infer

$$\Delta \hat{r} \Delta \hat{s} > \tfrac{1}{2} |(r, s)| = \tfrac{1}{2} |(\hat{r}, \hat{s})|. \tag{11.33}$$

In other words, no matter how accurately the values of $A$ and $B$ are known for the first apparatus, the second apparatus cannot succeed in forcing the system into a state in which the values of $\hat{r}$ and $\hat{s}$ are known to accuracies better than that allowed by the HRS uncertainty relation. In one sense this is, of course, a trivial result, since the uncertainty relation is an abstract property of operators and vectors in Hilbert space, which must always hold. However, it shows once again the consistency of the quantum-mechanical formalism with the general theory of measurement.

* * *

The writing of these notes was made possible through the time and facilities extended to me by the Faculté des Sciences of the University of Paris and the Società Italiana di Fisica in Varenna. The arrangements were made by Drs. Y. Choquet-Bruhat and B. d'Espagnat, to whom I wish to express my warmest thanks.

## Appendix A

**Properties of the relative frequency function.**

The first property is elementary:

$$\text{(A.1)} \qquad \sum_{s_1 \dots s_N} f(s;\, s_1 \dots s_N) w_{s_1} \dots w_{s_N} = \frac{1}{N} \sum_{s_1 \dots s_N} (\delta_{ss_1} + \dots + \delta_{ss_N}) w_{s_1} \dots w_{s_N} = \frac{1}{N}(w_s + \dots + w_s) = w_s\,.$$

The proof of the second property requires a simple induction:

$$\text{(A.2)} \qquad \sum_{s_1 \dots s_N} [f(s;\, s_1 \dots s_N) - w_s]^2 w_{s_1} \dots w_{s_N}$$

$$= \sum_{s_1 \dots s_N} \left[\frac{1}{N^2}(\delta_{ss_1} + \dots + \delta_{ss_N})^2 - \frac{2}{N}(\delta_{ss_1} + \dots + \delta_{ss_N}) w_s + w_s^2\right] w_{s_1} \dots w_{s_N}$$

$$= \frac{1}{N^2} \sum_{s_1 \dots s_N} (\delta_{ss_1} + \dots + \delta_{ss_N})^2 w_{s_1} \dots w_{s_N} - w_s^2$$

$$= \frac{1}{N^2} \sum_{s_1 \dots s_N} [(\delta_{ss_1} + \dots + \delta_{ss_{N-1}})^2 + 2(\delta_{ss_1} + \dots + \delta_{ss_{N-1}})\delta_{ss_N} + \delta_{ss_N}^2] w_{s_1} \dots w_{s_N} - w_s^2$$

$$= \frac{1}{N^2}\Big[\sum_{s_1 \dots s_N} (\delta_{ss_1} + \dots + \delta_{ss_{N-1}})^2 + 2(N-1) w_s^2 + w_s\Big] - w_s^2$$

$$= \frac{1}{N^2}[2(1 + 2 + \dots + N-1) w_s^2 + N w_s] - w_s^2$$

$$= \frac{1}{N^2} N(N-1) w_s^2 + \frac{1}{N} w_s - w_s^2 = \frac{1}{N} w_s (1 - w_s)\,.$$

## Appendix B

**Identities satisfied by the Green's functions.**

To avoid misunderstandings in the derivation of eqs. (6.8) and (6.9) it is helpful to take note of the fact that an expression like $f^i S_{,ij} g^j$, where $f$ and $g$ are arbitrary functions, may be ambiguous. Unless the intersection of the supports of $f$ and $g$ occupies only a finite domain of space-time the associative law of multiplication may fail, and $(f^i S_{,ij})g^j$ may not be equal to $f^i(S_{,ij}g^j)$. This is because an integration by parts is generally necessary to pass from one form to the other. In the following equations, involving Green's functions, the student may easily verify that, owing to the kinematical structure of the Green's functions, the associative law does in fact hold.

Equation (6.8) follows immediately from (6.2) by writing:

$$\text{(B.1)}\quad 0 = -G^{+ki}(S_{,kl} - S_{,lk})G^{-lj} = -G^{+ki}(S_{,kl}G^{-lj}) + (S_{,lk}G^{+ki})G^{-lj} = G^{+ji} - G^{-ij}.$$

Equation (6.9) may be derived by first asking what the change in the retarded Green's function $G^-$ would be if the action suffered an infinitesimal change $\delta S$. Denoting this change by $\delta G^-$ we obtain immediately from eq. (6.6) the following variational equation:

$$\text{(B.2}a\text{)}\qquad \delta S_{,ik}G^{-kj} + S_{,ik}\delta G^{-kj} = 0\,,$$

or

$$\text{(B.2}b\text{)}\qquad S_{,ik}\,\delta G^{-kj} = -\delta S_{,ik}G^{k-j}\,.$$

$\delta G^-$ is seen to satisfy an inhomogeneous linear differential equation similar to the equation of small disturbances. The solution of this equation that incorporates the boundary conditions necessary to maintain the integrity of the kinematical structure of $G^-$ is

$$\text{(B.3)}\qquad \delta G^{-ij} = G^{-ik}\delta S_{,kl}G^{-lj}\,.$$

This may be combined with the symmetry law (B.1) to yield

$$\text{(B.4)}\qquad \delta G^{\pm ij} = G^{\pm ik}G^{\mp jl}\delta S_{,kl}\,.$$

Equation (6.9) is essentially just a special case of this.

## Appendix C

### The Poisson-Jacobi identity.

Let $r_1, r_2, r_3$ be any three observables and let $\varepsilon_{\alpha\beta\gamma}$ be the antisymmetric permutation symbol in three dimensions. Then

$$\text{(C.1)}\qquad \varepsilon_{\alpha\beta\gamma}\big(r_\alpha(r_\beta, r_\gamma)\big) = \varepsilon_{\alpha\beta\gamma}r_{\alpha,i}\tilde{G}^{il}(r_{\beta,j}\tilde{G}^{jk}r_{\gamma,k})_{,l} =$$
$$= -\varepsilon_{\alpha\beta\gamma}r_{\alpha,i}r_{\gamma,k}\tilde{G}^{il}\tilde{G}^{kj}r_{\beta,jl} + \varepsilon_{\alpha\beta\gamma}r_{\alpha,i}r_{\beta,j}\tilde{G}^{il}\tilde{G}^{jk}r_{\gamma,kl} +$$
$$+ \varepsilon_{\alpha\beta\gamma}r_{\alpha,i}r_{\beta,j}r_{\gamma,k}(G^{+il} - G^{-il})(G^{+jm}G^{-kn} - G^{-jm}G^{+kn})S_{,lmn};$$

in which use has been made of eqs. (6.9) and (7.4). The factor multiplying $\varepsilon_{\alpha\beta\gamma}$ in the first term is symmetric in $\alpha$ and $\gamma$ while that multiplying $\varepsilon_{\alpha\beta\gamma}$ in the second term is symmetric in $\alpha$ and $\beta$. These two terms therefore vanish. In the third term the factor $\varepsilon_{\alpha\beta\gamma}\, r_{\alpha,i} r_{\beta,j} r_{\gamma,k}$ is invariant under cyclic permutation of $i, j, k$. Therefore, if the factors involving the Green's functions are multiplied out, the resulting four terms may be subjected to independent cyclic permutations of $i, j, k$ without affecting the value of the expression. Because of the symmetry of $S_{,lmn}$ the indices $l, m, n$ may be subjected to corresponding permutations. The following replacement is therefore permitted:

$$\begin{array}{cccc} G^{+il}G^{+jm}G^{-kn} - & G^{+il}G^{-jm}G^{+kn} - & G^{-il}G^{+jm}G^{-kn} + & G^{-il}G^{-jm}G^{+kn}\,, \\ \downarrow & \downarrow & \downarrow & \downarrow \\ G^{+jm}G^{+kn}G^{-il} - & G^{+kn}G^{-il}G^{+jm} - & G^{-kn}G^{+il}G^{-jm} + & G^{-jm}G^{-kn}G^{+il}\,. \end{array}$$

The resulting term by term cancellation shows that the third term in (C.1) likewise vanishes, and hence that the Poisson-Jacoby identity is satisfied.

## REFERENCES


[1] H. Everett III: *Rev. Mod. Phys.*, 29, 454 (1957).

[2] N. Bohr and L. Rosenfeld: *Kgl. Danske Videnskab. Selskab, Mat.-fys. Med.*, 12, No. 8 (1933).

[3] J. A. Wheeler: *Rev. Mod. Phys.*, 29, 463 (1957).

[4] J. von Neumann: *Mathematical Foundations of Quantum Mechanics* (Princeton N. J., 1955).

[5] N. Bohr: as quoted by A. Petersen in the *Bulletin of the Atomic Scientists*, 19, No. 7, 8 (1963).

[6] W. Heisenberg: *The Physicist's Conception of Nature*, translated by A. J. Pomerans (London, 1958).

[7] R. N. Graham: Ph. D. thesis, University of North Carolina at Chapel Hill (1971).

[8] B. S. De Witt: *The Everett-Wheeler Interpretation of Quantum Mechanics*, in *Battelle Rencontres, 1967 Lectures in Mathematics and Physics*, edited by C. De Witt and J. A. Wheeler (New York, 1968).

[9] J. B. Hartle: *Am. Journ. Phys.*, 36, 704 (1958).

[10] R. E. Peierls: *Proc. Roy Soc.*, A 214, 143 (1952).

Reprinted from AMERICAN JOURNAL OF PHYSICS, Vol. 37, No. 12, 1212-1220, December 1969
Printed in U. S. A.

# On the Interpretation of Measurement within the Quantum Theory*

LEON N. COOPER AND DEBORAH VAN VECHTEN
*Brown University, Providence, Rhode Island 02912*
(Received 26 May 1969)

An interpretation of the process of measurement is proposed which can be placed wholly within the quantum theory. The entire system including the apparatus and even the mind of the observer can be considered to develop according to the Schrödinger equation. No separation, in principle, of the observer and the observed is necessary; nor is it necessary to introduce either the type I process of von Neumann or wave function reduction.

## INTRODUCTION

Although the structure of the quantum theory in the opinion of almost all physicists is free from contradiction, questions about the consistency of its interpretation have been and continue to be posed. The view expressed in most texts and taught in many classes derives from the work of von Neumann[1] in the early thirties; it implies what is close to a Cartesian dualism dividing mind and body which, though consistent (and perhaps even respectable in the 17th century[2,3]), seems somewhat of an anachronism at present.[4,5]

Von Neumann proposed that the interpretation of measurement—or the means by which we come to know that something has happened—requires a process which does not develop according to the Schrödinger equation. He says:[6] "We therefore have two fundamentally different types of interventions which can occur in a system $S$ or in an ensemble $(S_1, \cdots, S_N)$. First, the arbitrary changes by measurements which are given by the formula

$$U \rightarrow U' = \sum_{n=1}^{\infty} (U\phi_n, \phi_n) P_{[\phi_n]}. \qquad \text{(I)}$$

Second, the automatic changes which occur with passage of time. These are given by the formula

$$U \rightarrow U_t = \exp\{-(2\pi i/h)tH\}\, U \exp\{(2\pi i/h)tH\}. \qquad \text{(II)}$$

Further:[7] ...we must always divide the world into two parts, the one being the observed system, the other the observer. In the former, we can follow up all physical processes (in principle at least) arbitrarily precisely. In the latter, this is meaningless."

Wigner recently has written:[8] "...one must conclude that the only known theory of measurement which has a solid foundation is the orthodox one and that this implies the dualistic theory concerning the changes of the state vector. It implies, in particular, the so-called reduction of the state vector."

* Supported in part by the Advanced Research Projects Agency, the National Science Foundation and the U. S. Atomic Energy Commission.

[1] J. von Neumann, *Mathematical Foundations of Quantum Mechanics*, transl. by Robert T. Beyer (Princeton University Press, Princeton, N. J., 1965).

[2] It was not accepted universally even then. Spinoza, for example, objected saying: "he [Descartes] accomplishes nothing beyond a display of the acuteness of his own great intellect."

[3] B. Spinoza, *The Ethics*, D. D. Runes, Ed. (Wisdom Library, division of The Philosophical Library, New York, 1957), p. 24.

[4] The number of papers on this subject is large and the individual contributions not always easy to understand. We have not made an exhaustive study of the literature and make no claim that every concept presented is written down for the first time; however we have never seen the entire matter discussed in this light. Our own primary references were Wigner's 1963 article, von Neumann's book, and what a somewhat reversible memory told us we had read and been taught over the years. We were directed to Prof. K. Gottfried's excellent discussion in his book on Quantum Mechanics (Ref. 5) somewhat later. There he has stated the relation between measurement and irreversibility in a very clear and elegant fashion. We would like to express our gratitude to Prof. H. P. Stapp for a very interesting correspondence and for bringing our attention to an article of Hugh Everett III, Rev. Mod. Phys. **29**, 454 (1957). Everett, whose views do not seem to be generally known, recognizes the necessity of retaining all branches of the wave-function; in this respect his ideas are quite similar to our own.

Just as Everett we retain all branches of the wave function. However, it is not the wave function itself which is put into correspondence with experience. Rather this correspondence is made via the amplitude 35. Thus there are amplitudes which give the probability for any particular sequence of events that might constitute an evolving world. These is nothing, however, which necessitates that more than one of these come to pass. (Last paragraph of footnote added November 1972.)

[5] K. Gottfried, *Quantum Mechanics*, (W. A. Benjamin, Inc., New York, 1966), vol. I, pp. 165–189.

[6] See Ref. 1, p. 351.

[7] See Ref. 1, p. 420.

[8] E. P. Wigner, Amer. J. Phys. **31**, 6 (1963), p. 12.

In the words of Spinoza[9] "They appear to conceive man to be situated in nature as a kingdom within a kingdom: for they believe he disturbs rather than follows nature's order. . . ."

In what follows we propose a reinterpretation of the quantum theory in which those processes called "measurement" or cognition are brought within the ordinary time development of the Schrödinger equation. Thus we have no need for the discontinuous process [Eq. (I)] of von Neumann. In this way perception or cognition are made a part of nature so that, again in the words of Spinoza,[10] "mind and body are one and the same thing. . . ."

## I. ANALYSIS OF AN INTERFERENCE PATTERN

Consider a traditional arrangement (Fig. 1) which results in the interference pattern due to the passage of an electron through a barrier with two slits. The wave function of a single electron localized in the region $R$, at the time $t_0$, moving to the right in the *horizontal* direction may be written

$$\psi_0(t_0). \tag{1}$$

This develops according to the Schrödinger equation so that at time $t_1$ the packet is passing through the slits and we might write

$$\Psi(t_1)=\alpha\psi_U+\beta\psi_L, \tag{2}$$

where $\psi_U(\psi_L)$ is the part of the packet that passes through the upper (lower) slit and is normalized. At time $t_2$ when the packets have reached the screen, the wave function can be written

$$\Psi(t_2)=\alpha\psi_U(t_2)+\beta\psi_L(t_2). \tag{3}$$

The probability amplitude that the electron arrive at the point $x_0$ on the screen is given by

$$\langle\Psi(t_2,x)\mid x_0\rangle=\alpha\langle\psi_U(t_2,x)\mid x_0\rangle +\beta\langle\psi_L(t_2,x)\mid x_0\rangle. \tag{4}$$

We cannot say whether the electron has "gone" through the upper or lower slit since the amplitude is a coherent sum of $\psi_U$ and $\psi_L$.

[9] See Ref. 3, p. 23.

[10] See Ref. 3, p. 27.

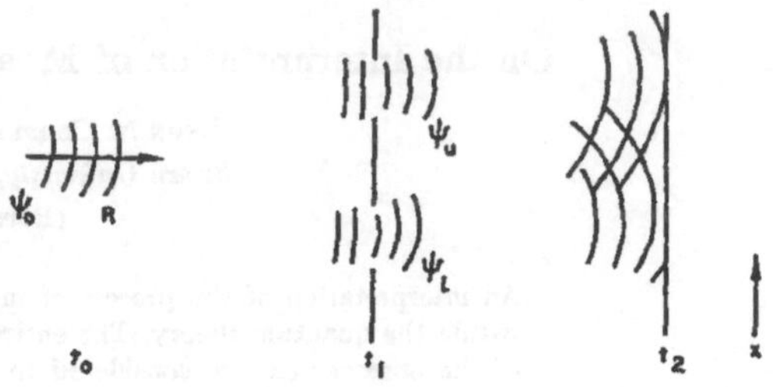


FIG. 1. Traditional arrangement resulting in an interference pattern.

In the double Stern–Gerlach experiment, discussed by both Bohm[11] and Wigner,[12] an electron prepared in a state which is an eigenfunction of $\sigma_z$ is acted upon by an inhomogeneous magnetic field in the $x$ direction so that

$\psi_U$ is correlated with spin $\uparrow$

and

$\psi_L$ is correlated with spin $\downarrow$.

We thus have at $t_1$

$$\Psi(t_1)=\alpha u_\uparrow\times\psi_U+\beta u_\downarrow\times\psi_L. \tag{5}$$

Is this to be regarded as a measurement of the spin state of the electron?

In the usual sense the answer must be no since at the screen at time $t_2$ the amplitude that the electron arrive at $x_0$ is

$$\alpha u_\uparrow\langle\psi_U\mid x_0\rangle+\beta u_\downarrow\langle\psi_L\mid x_0\rangle. \tag{6}$$

Its wave function is thus (as expected) a superposition of the two spin states $u_\uparrow$ and $u_\downarrow$. This is made more graphic if we recombine the spatial wave packets (by the introduction of a current which reverses the effect of the inhomogeneous magnetic field) in such a way that

$$\alpha\psi_U(t_1)+\beta\psi_L(t_1)\rightarrow\psi_0(t_2), \tag{7}$$

where $\psi_0$ is the original wave packet. The wave function would then be

$$\Psi(t_2,x)=\psi_0(x)(u_\uparrow+u_\downarrow), \tag{8}$$

and the electron is again in an eigenstate of $\sigma_x$.

[11] D. Bohm, *Quantum Theory* (Prentice-Hall, Inc., Englewood Cliffs, N. J., 1951), Chap. 22.

[12] See Ref. 8, p. 10.

## II. MEASUREMENT

What then constitutes a measurement? To discover whether the electron has passed by the upper or lower path, we might place a detector at each slit. Call a detector $A \equiv (a_U \times a_L)$. The wave function of the entire system (electron+detector) is then written

$$\Psi = \psi \times A. \tag{9}$$

At $t_1$ the electron can interact with the detector and this interaction puts the detector system into the states

$$A_U \equiv (a_U^+ \times a_L) \tag{10}$$

and

$$A_L \equiv (a_U \times a_L^+).$$

The meaning of the latter two equations is that the electron interacts with the upper detector and not with the lower, or with the lower detector and not with the upper.[13]

We assume in what follows that the state represented by $A_U$ is orthogonal to that represented by $A_L$ so that for all time

$$\langle A_U \mid A_L \rangle = 0. \tag{11}$$

The wave function

$$\Psi(t_0) = \psi_0(t_0) \times A(t_0) \tag{12}$$

can now develop in time to give at $t_1$

$$\Psi(t_1) = \alpha\psi_U \times A + \beta\psi_L \times A. \tag{13}$$

If we assume that the detectors are so designed that there must be an interaction in order for the wave function to pass through the slits, we obtain at $t_1+\delta$ the correlations

$$\psi_U \times A \rightarrow \psi_U \times A_U,$$

$$\psi_L \times A \rightarrow \psi_L \times A_L. \tag{14}$$

Thus all together we have

$$\Psi(t_1+\delta) = \alpha\psi_U \times A_U + \beta\psi_L \times A_L. \tag{15}$$

This wave function develops further so that at $t_2$ we have

$$\Psi(t_2) = \alpha\psi_U(t_2) \times A_U(t_2) + \beta\psi_L(t_2) \times A_L(t_2). \tag{16}$$

Since $\langle A_U \mid A_L \rangle = 0$ it follows that the probability amplitude that the electron arrive at $x_0$ and that the detector read $A_U$ (the electron has gone through the upper slit) is

$$\langle \Psi(t_2) \mid x_0 A_U \rangle = \alpha \langle \psi_U(t_2) \mid x_0 \rangle \tag{17}$$

while the amplitude that the electron arrive at $x_0$ and that the detector read $A_L$ (the electron has gone through the lower slit) is

$$\langle \Psi(t_2) \mid x_0 A_L \rangle = \beta \langle \psi_L(t_2) \mid x_0 \rangle \tag{18}$$

and of course there is no interference.

Therefore unless $A_U$ and $A_L$ can develop into states which are not orthogonal to each other, or if one requests a matrix element which is a superposition of $A_U+A_L$ (that is a superposition of two orthogonal macroscopic states which is not normally done) there can be no further interference between $\psi_U$ and $\psi_L$ and the wave function [Eq. (16)] is not distinguishable from a mixture.

Whether $A_U$ and $A_L$ will develop into states which are not orthogonal to each other depends upon whether the states $A_U$ and $A_L$ are reversible. In a microscopic sense, of course, all states are reversible. But for large systems in the thermodynamic sense certain states are not.

It is this that distinguishes the double Stern–Gerlach experiment from what we usually understand as a measurement. For the former there exists an easily realizable interaction (that produced by a current loop for example) which will reverse the effect of the original inhomogeneous magnetic field and thus bring $\psi_U$ and $\psi_L$ back to $\psi_0$. The interaction in what is usually understood as a detector (e.g., an interaction in a photographic plate) cannot usually be made to reverse itself. Therefore, the states $A_U$ and $A_L$ do not, in the normal course of events, develop into states which are not orthogonal to each other. We take the point of view that the existence of interactions which are macroscopically or thermodynamically irreversible is what removes the possibility of future interference and makes a coherent wave function indistinguishable from a mixture. Therefore, a measurement or the prepara-

---

[13] This of course is an hypothesis in addition to the Schrödinger equation. The electron is indivisible. In quantum field theory we would write:

$$[(\psi_U^*\psi_U)(\psi_L^*\psi_L), \psi^*(x)] = (\psi_U^*\psi_U)\psi_L^*\delta(x-x_L) + \psi_U^*(\psi_L^*\psi_L)\delta(x-x_U).$$

tion of a state is in fact the interaction of a system to be observed or prepared with another system which can be put into a state that is irreversible for reasons of entropy, thus eliminating the possibility of future interference. This second system (often called the apparatus or even the observer) is usually thought of as being classical. But this from our point of view is not necessary. Any quantum system (a nucleus which can undergo fission for example) which can suffer a large increase in entropy due to the interaction will do as well.

### III. COGNITION

We now propose that the mind is a system which, though large and capable of essentially irreversible changes, can be described within the Schrödinger equation. Thus a mind coupled (via the usual sensory organs) to the apparatus $A\equiv(a_U\times a_L)$ of Eq. (10) is to be denoted by a wave function of the form

$$\Psi=\psi\times A\times M. \quad (19)$$

As the wave function develops in time we have eventually

$$\Psi=\alpha\psi_U\times A_U\times M_U+\beta\psi_L\times A_L\times M_L. \quad (20)$$

It is important to recall that just as

$$A_U\equiv(a_U^+\times a_L)$$

and

$$A_L\equiv(a_U\times a_L^+), \quad (10)$$

so also $M_U$ means cognition of a registration in the upper detector and *no* registration in the lower while $M_L$ means the reverse.

Thus finally the wave function is a linear superposition of the two states

$$\psi_U\times A_U\times M_U$$

and

$$\psi_L\times A_L\times M_L. \quad (21)$$

Every instrument coupled to $U$ or $L$ in the future registers invariably $U$ or $L$ guaranteeing a complete correlation between the state of the interacting instrument and the branch of the wave function with which it is associated.

If we now agree that one of the instruments coupled to the wave function (say $A$ or $M$ above) interacts with the system in an irreversible manner in the sense just discussed, then once this interaction occurs, the two branches of the wave function can no longer interfere and we can assert just as classically that the system is either in state $U$ or $L$.

We have no difficulty describing a system of instruments, other minds (or even our own mind if we are speaking of the future) and stating that the system (instruments, other minds) is, will be, or was, either in state $U$ or $L$, each with the probability amplitude given by the proper inner product.

It is the process of discovery or cognition that is difficult to describe. We can say of our own mind that it will be the state $U$ or $L$ with probability $|\alpha|^2$ or $|\beta|^2$ at the future time $t_1$. But to say that our own mind was in either of the states $U$ or $L$ in the past means that we have forgotten; to say that right now our mind is either in the state $U$ or $L$ means that we do not know the state of our own mind. When one speaks of cognition it is tautologically implied that one knows the state of one's own mind—that one has discovered which of the various possibilities in actual fact has come to pass.

In classical physics an ensemble is sometimes said to be reduced to a pure state

$$\rho_U\times A_U+\rho_L\times A_L\rightarrow\rho_U\times A_U, \quad (22)$$

when a system coupled to the observer, $M^e$ for example, registers $M^e{}_U$. The observer then knows what was in principle always knowable. The transition implied by Eq. (22) is not describable by a Hamiltonian since in effect it represents a (retrospective) change in initial condition. The state determined for all time by the Hamiltonian and by the initial condition was never fully determined, since the initial conditions were never completely known. When an observation is made, the state of the system (past, present and future) is determined.

If the entire process including the mind were put into the equations of motion, then we might write to begin

$$\rho_U\times A\times M_{(\rho_U+\rho_L)\times A}, \quad (23)$$

which means that the system is in the state $U$ but the mind is aware only of the ensemble $U+L$. As

FIG. 2. A reversible simple mind.

time passes this becomes

$$\rho_U \times A_U \times M_{\rho_U \times A_U}. \quad (24)$$

Thus in fact we have not gone from

$$\rho_U + \rho_L \rightarrow \rho_U, \quad (25)$$

but rather from

$$M_{(\rho_U+\rho_L)\times A} \rightarrow M_{\rho_U \times A_U}. \quad (26)$$

This represents a change in the state of excitation of the mind which, with its sensory coupling, can be assumed to be describable by a Hamiltonian within the compass of classical physics.

In quantum physics the situation is in some respects similar, but in a fundamental sense it is very different. If the wave function of a quantum system coupled to instruments, minds, etc. at $t_0$ is

$$\Psi(t_0) = \alpha\psi_U \times A \times M \times \cdots M^e + \beta\psi_L \times A \times M \times \cdots M^e, \quad (27)$$

and develops at some later time into

$$\Psi = \alpha\psi_U \times A_U \times M_U \times \cdots M^e{}_U + \beta\psi_L \times A_L \times M_L \times \cdots M^e{}_L, \quad (28)$$

and if we then regard this system ourselves, we "discover" for example that it is in the state $U$. This according to tradition "reduces" the wave function in close analogy with the classical reduction of an ensemble (the discontinuous process of von Neumann not describable by a unitary transformation) so that upon cognition,

$$\Psi = \alpha\psi_U \times A_U \times \cdots M^e{}_U + \beta\psi_L \times A_L \times \cdots M^e{}_L \quad (29)$$

suddenly becomes

$$\Psi = \psi_U \times A_U \times M_U \times \cdots M^e{}_U, \quad (30)$$

and is renormalized. This reduction, however, is not incorrect only in the case in which at least one of the devices (mind included) within the correlated system cannot reverse itself. Were such total reversal to occur, the upper and lower correlated branches would again interfere, and by reducing the wave function we would be throwing away a branch needed to produce the resulting interference pattern. Thus the practice of reducing the wave function upon coming to "know" the state of the system is either a manner of speaking or incorrect.

We can perhaps make this more graphic by considering what might be called a reversible mind. For simplicity, we assume that the mind may register only two states, $U$ and $L$. (As a concrete example, consider a 2-level atom and a photon of the right energy, Fig. 2.) Transitions between the two states are possible so that if there are no irreversible devices coupled to the system, it oscillates between $U$ and $L$. Thus the state $M_U(t_1)$ does not preclude the possibility that this could develop into $M_L(t_1+\delta)$. In this process, of course, the mind would retain no memory of its previous state. In this kind of "mind reversal" the two branches $U$ and $L$ would be expected to interfere just as in the Stern–Gerlach experiment with the introduction of the current loop. Thus a "reduction" of the wave function which discards one of the branches would be incorrect.

From this point of view the concept of "knowing something" has been introduced to order a world in which "mind reversals" do not normally occur. If "mind reversal" was a relatively frequent event, "knowledge" or "memory" would not exist in the usual sense and the concept of "knowing something" would not be likely to have been introduced in the same way.

## IV. HOW IS SOMETHING KNOWN?

How then is something known? A wave function which is a superposition of various amplitudes does not contain any information which indicates in which amplitude the system "really" is. [The type I process of von Neumann (wave function reduction) is designed precisely to put this information into the wave function.] We therefore must find some entity in the theory which corresponds to the seemingly evident fact that knowledge is possible.

When we are aware at the time $t_1$ that we ($M^e$) are in the state $M^e{}_U$, we no longer can require that the wave function have the form

$$\Psi(t_1)=\psi_U\times A_U\times B_U\times\cdots M^e{}_U(t_1), \quad (31)$$

although if it has the above form we will find ourselves in the state $M^e{}_U$. Rather what we require when we say that we "know" a system is in the state $U$ is that there will be agreement between ourselves, other observers and detectors (all systems coupled to us) that the system is in $U$. At any one time this agreement is contained in the matrix element

$$\langle\Psi(t_1)\mid\psi_U\times A_U\times B_U\times\cdots M^e{}_U(t_1)\rangle=\alpha, \quad (32)$$

which is the probability amplitude for a simultaneous observation of $\psi_U$, $A_U\cdots M^e{}_U$ at $t_1$. The probability amplitude for disagreement among systems or observers—the probability amplitude for a simultaneous observation of say $\psi_U$, $A_U$, $B_L\cdots M^e{}_U$ at $t_1$ is

$$\langle\Psi(t_1)\mid\psi_U\times A_U\times B_L\times\cdots M^e{}_U(t_1)\rangle=0. \quad (33)$$

The correlation among systems or observers on the various branches of the wave function makes a matrix element such as Eq. (33) zero. Because of this, our knowledge that our own mind is in some state need not be reflected in the wave function. Rather, it is expressed in the way we pose the question. That a system is in the state $U$ is equivalent to the statement that all coupled "good" systems and sane minds will agree that the system is in the state $U$ and this agreement is contained in Eq. (32). The probability amplitude for disagreement [Eq. (33)] is zero.

If we have a system of coupled detectors and minds which develop in time, the wave function splits and branches continuously. We must therefore construct some entity that corresponds to our experience that things happen at particular times and that we can know and agree with one another that they happened at those times.

This entity, we propose, is the scalar product defined below which gives the conditional probability amplitude for

$$Z_n(t_z')\text{ given }Y_m(t_y')\text{ given }\cdots$$

$$\text{given }C_k(t_c')\text{ given }B_j(t_b')\text{ given }A_i(t_a'), \quad (34)$$

where $t_z'>t_y'>\cdots t_c'>t_b'>t_a'$. This conditional amplitude we denote by

$$\mathcal{A}(A_i(t_a')\cdots Z_n(t_z')); \quad (35)$$

we assume that the various detectors (which now may include minds) are sufficiently decoupled so that the wave function can be written as a product, and that the Hamiltonian (other perhaps than for interactions over a very brief period) can be written as a direct sum. Excluding the times $t_a$, $t_b$, $t_c$, $\cdots$, $t_z$ the Hamiltonian for the entire system is written

$$H=H_A+H_B+\cdots H_Z. \quad (36)$$

At $t_a$ there is a possible interaction $H_{OA}$; at $t_b$ there is a possible interaction $H_{AB}$ etc. The interaction $H_{AB}$, for example, acts at the time $t_b$ in such a way that

$$A_U(t)B(t)\rightarrow A_U(t_b)B_U(t_b)$$

$$A_L(t)B(t)\rightarrow A_L(t_b)B_L(t_b). \quad (37)$$

At $t_c$ a similar possible interaction $H_{BC}$ takes $B_{U(L)}C\rightarrow B_{U(L)}C_{U(L)}$, and so on. A time $t_i'$ is defined to be any time other than the instants at which the detectors are changing their state (a process thought of as being short compared with the times over which the entire system develops). Thus if for example

$$t_i<t_i'<t_{i+1}$$

at the time $t_i'$ all of the systems $AB\cdots I$ have registered $U$ or $L$ while the systems $J\cdots Z$ have not yet registered.

The conditional amplitude [Eq. (35)] we now define as

$$\begin{aligned}\mathcal{A}(A_i(t_a')\cdots Z_n(t_z'))\equiv\langle\Psi(t_z')\mid\\ \times\exp\{-i[(H_A+H_{OA})(t_z'-t_a')\\ +(H_B+H_{AB})(t_z'-t_b')+\cdots\\ +(H_Y+H_{XY})(t_z'-t_y')]\}A_i(t_a')\\ \times B_j(t_b')\times\cdots Y_m(t_y')\times Z_n(t_z')\rangle. \quad (38)\end{aligned}$$

If the times $t_a'$, $t_b'\cdots$ all occur after the detectors $A, B\cdots$ have registered (i.e., $t_a'>t_a$, $t_b'>t_b\cdots t_z'>t_z$), the conditional amplitude takes the relatively simple form

$$\begin{aligned}&\mathcal{A}(A_i(t_a')B_j(t_b')\cdots Z_n(t_z'))\\ &=\langle\Psi(t_z')\mid\exp[-iH_A(t_z'-t_a')]A_i(t_a')\\ &\times\exp[-iH_B(t_z'-t_b')]\\ &\times B_j(t_b')\cdots\exp[-iH_y(t_z'-t_y')]Y_m(t_y')Z_n(t_z')\rangle.\end{aligned} \quad (39)$$

As an example consider the conditional amplitude

$$\mathfrak{a}[A_U(t_a')B_j(t_b')C_k(t_c')] = \langle\Psi(t_c') \mid \exp\{-i[(H_A+H_{OA})(t_c'-t_a') + (H_B+H_{AB})(t_c'-t_b')]\}A_U(t_a')B_j(t_b')C_k(t_c')\rangle. \quad (40)$$

Let us assume that $t_a<t_a'<t_b$ and $t_c'>t_c$ so that the detectors $A$ and $C$ have registered.

(1) If $t_b'>t_b$, the inner product becomes

$$\langle\Psi(t_c') \mid \exp-i[H_A(t_c'-t_a')]A_U(t_a') \times\exp-i[H_B(t_c'-t_b')]B_j(t_b')C_k(t_c')\rangle = \langle\Psi(t_c') \mid A_U(t_c')B_j(t_c')C_k(t_c')\rangle. \quad (41)$$

An essential point in the argument is just that the state recognizable as $A_U$ at $t_a'$ develops into a state still recognizable as $A_U$ at $t_c'$. (The blackened grain on a photographic plate remains recognizable as a blackened grain.)

The wave function at $t_c'$ is

$$\Psi(t_c') = (\alpha\psi_U\times A_U\times B_U\times C_U +\beta\psi_L\times A_L\times B_L\times C_L)\times D\times E\cdots.$$

Therefore the conditional amplitude is zero unless $j$ and $k$ are both $U$. Thus given that $A$ has registered $A_U$ at $t_a'$, we can conclude that $B$ will register $B_U$ at $t_b'$ and $C$ will register $C_U$ at $t_c'$.

(2) If $t_b'<t_b$ the detector $B$ has not yet registered ($B_j=B$) and the inner product becomes

$$\langle\Psi(t_c') \mid \exp\{-i[H_A(t_c'-t_a') +(H_B+H_{AB})(t_c'-t_b')]\}A_U(t_a')B(t_b')C_k(t_c')\rangle; \quad (43)$$

at $t_b$, $A_UB\to A_UB_U$ due to the interaction $H_{AB}$; at the later time $t_c'$ the scalar product becomes

$$\langle\psi(t_c') \mid A_U(t_c')B_U(t_c')C_k(t_c')\rangle, \quad (44)$$

which equals zero unless $k=U$.

Thus for $t'_a>t_a$, $t'_b<t_b$, $t'_c>t_c$ (and as before $t_a'<t_b'<t_c'$) we find a nonzero amplitude for $A_U$, $B$ unregistered and $C_U$. The interaction $H_{AB}$ converts $B$ unregistered at $t_b'$ into $B_U$ for times later than $t_b$.

We can now say that something occurred (e.g., a measurement) after the first irreversible interaction. For after $t_a$ (when $A$ has registered) we can say

$$\text{given } A_U(t_a')\to B_U(t_b')\to C_U(t_c') \text{ etc.,} \quad \text{if } t_b'>t_b \text{ and } t_c'>t_c$$

(if the detectors have registered). Or

$$\text{given } A_L(t_a')\to B_L(t_b')\to C_L(t_c') \text{ etc.}$$

Before the irreversible interaction we could not say this because $\psi_U(t_1)$ might develop into $\psi_L(t_2)$. It is only with an irreversible interaction that we can be sure that the entire sequence of correlations will follow. We can therefore say that something happens when no interaction (in the usual sense) can reverse the situation: when given $A_U(t_a')$ implies that we will not find $A_L(t_b')$.

Our mind in this respect is like any other irreversible system. We can say that given $A_U(t_a')$ our mind will register $M'_U(t_m')$ as the conditional amplitude

$$\mathfrak{a}[A_U(t_a')\cdots M'_L(t_m')]=0. \quad (45)$$

Therefore we can say that if $U$ occurred $I$ will see it as $U$ or if $I$ see it as $U$ it was $U$. There is nothing special about cognition. It is one act in a correlated chain.[14]

To measure or observe the same quantity more than once, we envisage several detectors coupled to the original system; say

$$S\times A\times B\times C\times\cdots Z, \quad (46)$$

in such a way that

$$S_U\to\begin{cases}A_U\\ B_U,\\ C_U\end{cases} \quad (47)$$

[14] That we perceive one possibility or another (say $U$ or $L$) even though the wave function is a superposition of both $U$ and $L$ is due presumably to the nature of that physiological system called mind. We have only to assume that the mind shares the property of the detector [Eq. (10)] of registering $U$ and not $L$ or $L$ and not $U$. The wave function may contain a superposition of $U$ and $L$ but there is no manifestation of this to anyone—including the mind described by the wave function—unless interference can occur.

which implies a coupling of the form

$$H_{SA},\, H_{SB},\, H_{SC}\cdots. \tag{48}$$

The conditional amplitude that $A$ be $A_i$ at $t_a'$, $B$ be $B_j$ at $t_b'$, $\cdots Z$ be $Z_n$ at $t_z'$ given that $S$ is $S_U$ at $t_0$ (i.e., that $S$ is observed to be in the states $ijk\cdots n$ at the times $t_a'\cdots t_z'$) is

$$\langle\Psi(t_z')\mid\exp\{-i[H_S(t_z'-t_0)+(H_A+H_{SA})(t_z'-t_a')$$
$$+(H_B+H_{SB})(t_z'-t_b')+\cdots(H_Y+H_{SY})(t_z'-t_y')]\}$$
$$\times S_U(t_0)A_i(t_a')\cdots Y_m(t_y')Z_n(t_z')\rangle, \tag{49}$$

and this by the discussion above is zero, if $S_U$ cannot reverse itself, unless $AB\cdots Z$ are either unregistered or in the state $U$ depending upon whether $t_i'$ is smaller or larger than $t_i$.

The persistence of our memory can be interpreted also as a result of the irreversibility of mental processes. When we observe a detector to have registered $U$ at some time $t_i$ we both remember this at later times and can verify by looking again that the detector has in fact registered $U$. How for example, do we ask for the amplitude that the memory be in the state $U$, given that it was in $U$ at some previous time? We imagine that the mind can be thought of as many detectors $A, B, C\cdots$ all coupled to external objects (via the sensory organs) and to an internal memory. By coupling $A, B, C\cdots$ with the memory (considered an irreversible system) or to some external apparatus (either denoted by $S$) we can precisely as in the manner of Eqs. (46)–(49) show that given $S_U$ at $t_0$ implies $A_U(t_a')B_U(t_b')$ etc. Thus we have agreement internally as to what we remember and what we see.[15]

If the mind were not thought to be irreversible, $M^{\circ}{}_U(t_0)$ could develop into $M^{\circ}{}_L(t)$, or to some other state not orthogonal to $M^{\circ}{}_L$. The essence of the assumption of irreversibility is that $M^{\circ}{}_U(t_0)$ develops into a state always recognizable as $M_U$ and always orthogonal to $M^{\circ}{}_L$. Thus the amplitude

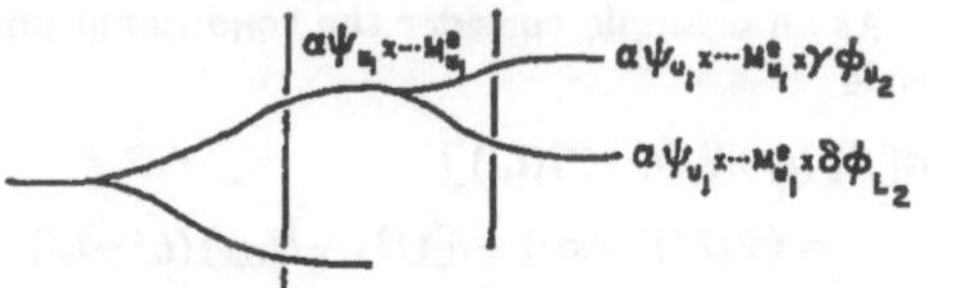


FIG. 3. A second double slit following the first.

that the mind be in the state $U$, given that we found ourselves in $U$ at some previous time, is always the same and always comes from the same branch of the wave function. The amplitude that we find $L$, given that we found $U$ previously, is zero since $M_L$ cannot develop from $M_U$.

How then do we interpret the non-zero amplitude

$$\langle\psi\mid\psi_L\times A_L\times\cdots M^{\circ}{}_L\rangle=\beta, \tag{50}$$

when we "know" that the system is in the state $U$? This amplitude we assert is in fact non-zero and corresponds to the amplitude that the system and all of the minds (including our own) are in the state $L$. (We don't usually ask for the amplitude that our mind be in the state $L$ if we know it to be in $U$). That this amplitude is nonzero has no effect on the other amplitude ($U$) if the two can never interfere. If, as in Fig. 3, we arrange another double slit so that the upper branch is split into two parts, one is concerned only with the conditional probability (given $M^{\circ}{}_U$ for the first set of slits) for the various possible outcomes of the second process. As long as there is no possibility of interference, and our questions are prefaced by "given $M^{\circ}{}_U$," we can discard the amplitude $\psi_L\times\cdots$ and renormalize $\psi_U\times\cdots$. Thus the so-called reduction is really a renormalization. If, however, the systems $A$, $M$ etc. can be reversed (brought back to a state in which the two amplitudes can interfere) then it is essential that the amplitude $\psi_L\times\cdots$ be retained so that the possibility of interference be preserved. We know something, therefore, because of the possibility of an infallible correlation between the state of our mind and other minds and systems, and not because the wave function of the world has no amplitudes for other possibilities. The other amplitudes reveal themselves only in mind

[15] When we see that a lamp is red our memory, $S$, with some irreversible change registers it as red, $S_R$. Later checking our memory continues to confirm that the lamp was registered as red. Each check is interpreted by the coupling of a detector $A$, $B$, $C\cdots$ with the memory and $S_R\rightarrow A_R$, $B_R$, $C_R$, etc., since $S_R$ is irreversible. Checking the lamp itself (if it is the kind of lamp that does not change color) reveals it to be red if it registered previously as red.

reversals and reversals of macroscopic systems that do not ordinarily occur.

## CONCLUSION

In conclusion we claim that the process of measurement according to the interpretation given above can be placed wholly within the quantum theory. The entire system including the apparatus and even the mind of the observer can be considered to develop according to the Schrödinger equation. No discontinuous processes need be assumed, nor is it necessary to introduce the idea of wave function reduction. The essential idea is that of the interaction of the system with a device which is irreversible in the thermodynamic sense and which eliminates as a practical matter the possibility of interference between the various branches into which the wave function is separated. This separation permits one to say that one is either on one branch or the other. The process of cognition (of being aware that something happened) is interpreted as corresponding to the certain correlation among the various branches of the wave function which allows the possibility of agreement among all observers and systems, and agreement in our own memory as to what happened. This certain correlation then allows one, as a matter of convenience, to discard the other branches and renormalize the first. This, however, is only a manner of speaking since if the interaction is reversible, the possibility of interference requires the retention of all of the branches of the wave function.

# THE MEASUREMENT OF RELATIVE FREQUENCY

Neill Graham

## 1. *Introduction*

In this paper we wish to explore the probability interpretation within the framework of the Everett interpretation of quantum mechanics [1, 2]. After a brief review of the Everett interpretation, including a critique of Everett's version of the probability interpretation, we propose a "two step" solution to the problem. In the first step an apparatus measures the relative frequency with which a given event occurs in a collection of independent, identically prepared systems. (Relative frequency is treated as an observable and is represented by a Hermitean operator. This approach to relative frequency was discussed by this author in his Ph.D. thesis [3]; a similar formulation was arrived at independently by J. B. Hartle [4].)

The second step of the proposed measurement process takes place after the apparatus has interacted with the systems in question and returned to thermal equilibrium. It is then "read" by a second apparatus (called the observer to prevent confusion). The important point is this: Since in the second step the system under observation is macroscopic, we can apply the powerful statistical techniques of Prosperi and Scotti [5], and of Daneri, Loinger, and Prosperi [6, 7] to a discussion of this observation. In particular, we can show that in the second measurement (but not the first) a relative frequency that agrees with the Born interpretation will be found in the overwhelming majority of the Everett worlds of the observer.

Let us begin our discussion by considering the measurement of an observable M with eigenstates $\psi_m$ and eigenvalues m. Experience tells us that a measurement of M on a system in the state $\psi$ will yield one of the eigenvalues m of M as a result. If the measurements do not disturb the system, and if M does not change spontaneously in time, then repeated measurements of M on the same system will yield the same eigenvalue. On the other hand, if a large number of independent measurements are made on identical systems all in the same state $\psi$, then the value m will be obtained with a relative frequency (number of occurrences divided by number of trials) close to $|\langle\psi_m,\psi\rangle|^2$.

(Some people have wondered whether repeated, nondisturbing measurements can actually be made on a single quantum system. The answer is clearly yes. Consider, for example, a scattering experiment where the scattered particle is detected in a cloud chamber. The wave function of this scattered particle is a spherical wave radiating out from the scattering center in all directions. If part of this wave is intercepted by the cloud chamber, straight line tracks will be observed in the chamber. Along a given track each atom performs an independent measurement of the direction of the outgoing particle (this being the direction of a line through the scattering center and the atom). After the ionization of the first atom on a track, each additional ionized atom, by virtue of lying on a line joining the first atom to the scattering center, confirms the direction of motion determined by the first atom. Thus the cloud chamber makes repeated measurements of the direction of motion of the same particle.)

Everett's great achievement was to give an explanation consistent with the Schrödinger equation of this apparent "reduction of the state vector." Let us look briefly at his approach. Consider a measuring apparatus in the initial state $\phi$. Then, when the system is in the eigenstate $\psi_m$, the initial state of the system and the apparatus is given by

$$\psi^i = \psi_m\phi \,. \tag{1.1}$$

Let U be the linear, unitary transformation that changes the initial state of the system and apparatus before the measurement into the final state afterwards. Then

$$\psi^f = U\psi^i \ . \tag{1.2}$$

If the measurement does not disturb the original state, $\psi_m$, of the system, then $\psi^f$ must have the form

$$\psi^f = \psi_m\phi[m] \ , \tag{1.3}$$

where $\phi[m]$ is the state of the apparatus when it has recorded the value m for M. (The different $\phi[m]$'s describe apparata with different "pointer readings.")

But what if the initial state, $\psi$, of the system is not an eigenstate of M? The result in this case is determined by the linearity of U. Suppose that $\psi$ has the expansion

$$\psi = \sum_m <\psi_m, \psi> \psi_m \ . \tag{1.4}$$

Then $\psi^i$ in turn is given by

$$\psi^i = \sum_m <\psi_m, \psi> \psi_m \phi \ , \tag{1.5}$$

and, because of the linearity of U, we have for $\psi^f$,

$$\begin{aligned} \psi^f = U\psi^i &= U(\sum_m <\psi_m, \psi> \psi_m \phi) \\ &= \sum_m <\psi_m, \psi> U(\psi_m \phi) \\ &= \sum_m <\psi_m, \psi> \psi_m \phi[m] \ . \end{aligned} \tag{1.6}$$

Thus $\psi^f$ is a superposition of states of the system and apparatus corresponding to different pointer readings of the apparatus.

According to Everett, this superposition describes a set of simultaneously existing worlds, one for each element of the superposition. In each world the apparatus has a unique pointer reading, the one described by the corresponding element, $\psi_m\phi[m]$, of the superposition. Everett further shows that, in each world, future nondisturbing measurements on the same system will give results consistent with the first measurement. Thus an apparent reduction of the state vector takes place in each world, yet the universal state vector describing all the worlds changes in accordance with the Schrödinger equation, linearly and deterministically.

Provided one is ready to accept the existence of multiple, simultaneously existing worlds, Everett's interpretation satisfactorily explains the apparent reduction of the state vector characteristic of quantum measurements. Such measurements, however, feature a second, equally important characteristic, the probability interpretation. In this case the explanation given by Everett is less convincing.

To see the problem, consider a collection of identical, independent systems, all in the same state $\psi$. These are the systems that might participate in some actual measurement of relative frequency, such as the particles that strike a photographic plate or pass through a particle detector. The state of this collection (viewed as a single system) is

$$\psi^N = \psi \ldots \psi \text{ (N terms)} . \tag{1.7}$$

Now consider an apparatus that will interact with each of these systems in turn. The initial state of the systems and the apparatus is

$$\psi^i = \psi^N \phi \tag{1.8}$$

$$= \sum_{m_1,\ldots,m_N} \langle \psi_{m_1}, \psi \rangle \ldots \langle \psi_{m_N}, \psi \rangle \psi_{m_1} \ldots \psi_{m_N} .$$

If $U$ is the linear operator describing the interaction, then we have

$$U\psi_{m_1} \ldots \psi_{m_N} \phi = \psi_{m_1} \ldots \psi_{m_N} \phi[m_1, \ldots, m_N] , \tag{1.9}$$

where $\phi[m_1, \ldots, m_N]$ is the state of an apparatus that has recorded the values $m_1, \ldots, m_N$. By the linearity of $U$ the final state of the systems and apparatus is

$$\psi^f = U\psi^i \tag{1.10}$$

$$= \sum_{m_1,\ldots,m_N} \langle\psi_{m_1},\psi\rangle \ldots \langle\psi_{m_N},\psi\rangle \psi_{m_1} \ldots \psi_{m_N} \phi[m_1,\ldots,m_N] \ .$$

According to the Everett interpretation this final state describes a set of worlds, one for each possible observed sequence, $m_1,\ldots,m_N$, of the outcomes of the measurements in question. Furthermore, every possible sequence of outcomes occurs in some world. If $\ell$ is some particular value of $m$, and $R$ the number of times $\ell$ occurs in a given sequence $m_1,\ldots,m_N$ (the number of "successes" in that sequence), then every possible value of $R$ from 0 to $N$ will be realized in some Everett world. There seems to be no connection between the actual relative frequency $\frac{R}{N}$ in a given world and the expected relative frequency $|\langle\psi_\ell,\psi\rangle|^2$.

To see this more clearly, assume that $m$ can take on $K$ possible values. Consider those sequences for which $\ell$ occurs at $R$ given positions. The number of such sequences is $(K-1)^{N-R}$, since there are $K-1$ possibilities for each position in which $\ell$ does *not* occur. If we multiply this by the number of ways of choosing the $R$ positions for $\ell$ to occur in, we get

$$\binom{N}{R}(K-1)^{N-R} \tag{1.11}$$

for the number of sequences with $R$ successes.

Table 1.1 gives this expression as a function of $R$ for the case $K=3$, $N=45$, and $P \equiv |\langle\psi_\ell,\psi\rangle|^2 = 0.8$. The values of $R$ consistent with the probability intérpretation are those near $NP = 36$. Now although there are $4.5 \times 10^{11}$ worlds for which $R = 36$ (and somewhat more for which $R$ is near that value), there are also $3.7 \times 10^{20}$ worlds in which $R = 15$ and $\frac{R}{N} = \frac{1}{3}$. This is over 800 million times as many as those for which the probability interpretation is satisfied.

In general, it can be shown that Expression (1.11) will have a sharp peak at $R = \frac{N}{K}$ (since it is proportional to a Bernoulli distribution with $p = \frac{1}{K}$ and $q = 1 - \frac{1}{K}$). Thus, except in the special case where $P = \frac{1}{K}$,

| R | NUMBER OF WORLDS WITH R SUCCESSES | MEASURE OF WORLDS WITH R SUCCESSES |
|---|---|---|
| 0 | 3.518437E+13 | 3.518440E−32 |
| 1 | 7.916484E+14 | 6.333192E−30 |
| 2 | 8.708132E+15 | 5.573209E−28 |
| 3 | 6.240828E+16 | 3.195306E−26 |
| 4 | 3.276435E+17 | 1.342029E−24 |
| 5 | 1.343338E+18 | 4.401854E−23 |
| 6 | 4.477794E+18 | 1.173828E−21 |
| 7 | 1.247386E+19 | 2.615960E−20 |
| 8 | 2.962541E+19 | 4.970324E−19 |
| 9 | 6.089667E+19 | 8.173422E−18 |
| 10 | 1.096140E+20 | 1.176973E−16 |
| 11 | 1.743859E+20 | 1.497965E−15 |
| 12 | 2.470467E+20 | 1.697694E−14 |
| 13 | 3.135593E+20 | 1.723812E−13 |
| 14 | 3.583535E+20 | 1.576057E−12 |
| 15 | 3.702986E+20 | 1.302873E−11 |
| 16 | 3.471549E+20 | 9.771550E−11 |
| 17 | 2.961028E+20 | 6.667646E−10 |
| 18 | 2.303021E+20 | 4.148757E− 9 |
| 19 | 1.636357E+20 | 2.358241E− 8 |
| 20 | 1.063632E+20 | 1.226286E− 7 |
| 21 | 6.331144E+19 | 5.839456E− 7 |
| 22 | 3.453352E+19 | 2.548125E− 6 |
| 23 | 1.726676E+19 | 1.019250E− 5 |
| 24 | 7.913931E+18 | 3.737250E− 5 |
| 25 | 3.323851E+18 | 1.255716E− 4 |
| 26 | 1.278404E+18 | 3.863742E− 4 |
| 27 | 4.498089E+17 | 1.087572E− 3 |
| 28 | 1.445814E+17 | 2.796613E− 3 |
| 29 | 4.237731E+16 | 6.557575E− 3 |
| 30 | 1.130062E+16 | .01398949 |
| 31 | 2.734020E+15 | .02707644 |
| 32 | 5.980669E+14 | .04738377 |
| 33 | 1.178011E+14 | .07466533 |
| 34 | 2.078842E+13 | .1054099 |
| 35 | 3.266752E+12 | .1325153 |
| 36 | 4.537156E+11 | .1472392 |
| 37 | 5.518163E+10 | .1432597 |
| 38 | 5.808592E+ 9 | .1206398 |
| 39 | 5.212839E+ 8 | .08661319 |
| 40 | 3.909629E+ 7 | .05196791 |
| 41 | 2.383920E+ 6 | .0253502 |
| 42 | 113520 | 9.657218E− 3 |
| 43 | 3960 | 2.695038E− 3 |
| 44 | 90 | 4.900068E− 4 |
| 45 | 1 | 4.355616E− 5 |

**Table 1.1. Number and measure of Everett worlds with R successes. Assumed are K=3, N=45, and P=0.8.**

R will differ from NP in the majority of Everett worlds. Only in a tiny minority of those worlds will the probability interpretation be even approximately true.

Everett attempts to escape from this dilemma by introducing a numerical weight for each world, which he defines to be

$$|\langle\psi_{m_1},\psi\rangle|^2 \dots |\langle\psi_{m_N},\psi\rangle|^2 \tag{1.12}$$

for the sequence $m_1,\dots,m_N$. To every set, S, of such sequences Everett assigns the measure

$$\mu(S) = \sum_{(m_1,\dots,m_N)\,\epsilon\, S} |\langle\psi_{m_1},\psi\rangle|^2 \dots |\langle\psi_{m_N},\psi\rangle|^2 \,. \tag{1.13}$$

Consider the set of all sequences with R successes. Suppose that these successes occur at fixed positions in the sequences. Then the measure of all such sequences is the product of N terms, one for each possible position in the sequence. For each position at which $\ell$ occurs, the corresponding term is $P = |\langle\psi_\ell,\psi\rangle|^2$. For each position at which a value other than $\ell$ occurs, the corresponding term is $Q = \sum_{m\neq\ell} |\langle\psi_m,\psi\rangle|^2 = 1-P$. Thus the measure of all those sequences in which $\ell$ occurs R times at fixed positions is $P^RQ^{N-R}$. This must be multiplied by the number of ways of choosing R fixed positions out of a totality of N. Doing this, the measure of those sequences having R successes is

$$\binom{N}{R} P^RQ^{N-R} \tag{1.14}$$

which is, of course, just the well known Bernoulli distribution. This distribution does indeed favor values of R near NP; in fact, for large N it approximates a Gaussian distribution with mean NP and standard deviation $\sqrt{NPQ}$. Since $\sqrt{NPQ} << NP$ for large N, this distribution will have a sharp peak around NP. Indeed, the measure of all sequences lying between $NP - 3\sqrt{NPQ}$ and $NP + 3\sqrt{NPQ}$ will be greater than 0.99 (for large N).

If we again refer to Table 1.1, we see that the measure of those worlds with R successes does indeed have a peak at $R = NP = 36$. But this is of little comfort when we observe the degree to which these worlds are in a numerical minority.

In short, we criticize Everett's interpretation on the grounds of insufficient motivation. Everett gives no connection between his measure and the actual operations involved in determining a relative frequency, no way in which the value of his measure can actually influence the reading of, say, a particle counter. Furthermore, it is extremely difficult to see what significance such a measure can have when its implications are completely contradicted by a simple count of the worlds involved, worlds that Everett's own work assures us must all be on the same footing.

(To be sure Everett argues that the measure defined by (1.13) is unique. But remember that Gleason [8] has shown that the probabilities defined by the Born interpretation, considered as a measure on a Hilbert space, are themselves unique. Nevertheless, this (hopefully) does not deter anyone from inquiring into the connection between those probabilities and experiments that measure relative frequency.)

It thus appears that, in the "one step" measurement we have described, any attempt to show that the probability interpretation holds in the majority of the resulting Everett worlds is doomed to failure. As mentioned earlier, we shall attempt to improve matters by considering instead a "two step" measurement, in which a macroscopic apparatus mediates between a microscopic system and a macroscopic observer. To better motivate the technical work that follows, we now give a brief outline of this approach.

To begin, define the relative frequency of $\ell$ in the sequence $m_1, \ldots, m_N$ by

$$f_\ell(m_1, \ldots, m_N) = \frac{1}{N} \sum_{i=1}^{N} \delta_{m_i \ell} \, . \tag{1.15}$$

Since relative frequency is an observable, on the same footing with any other observable in quantum mechanics, we may associate with it a Hermitean operator. We define this operator, $F_\ell$, in the obvious way:

its eigenstates shall be the states for which relative frequency is well defined, and its eigenvalues shall be the corresponding values of that relative frequency. Thus

$$F_\ell \psi_{m_1} \cdots \psi_{m_N} = f_\ell(m_1, \ldots, m_N) \psi_{m_1} \cdots \psi_{m_N} . \tag{1.16}$$

Thus $F_\ell$ is defined on a Hilbert space spanned by

$$\psi_{m_1},\ \psi_{m_1}\psi_{m_2},\ \psi_{m_1}\psi_{m_2}\psi_{m_3}, \ldots .$$

In order to state compactly the properties of $F_\ell$ that are of interest, it is useful to have the following definitions

$$\begin{aligned} <X>_\psi &\equiv <\psi, X\psi> , \\ \Delta_\psi X &\equiv \{<(X - <X>_\psi)^2>\}^{\frac{1}{2}} . \end{aligned} \tag{1.17}$$

These are, of course, introduced as mathematical abbreviations only. We cannot assume $<X>_\psi$ and $\Delta_\psi X$ to be the average and standard deviation of a set of measurements until the probability interpretation has in fact been established.

With the above definitions we can prove the following:

$$\begin{aligned} <F_\ell>_{\psi^N} &= |<\psi_\ell, \psi>|^2 , \\ \Delta_{\psi^N} F_\ell &= \left\{ \frac{|<\psi_\ell,\psi>|^2 (1 - |<\psi_\ell,\psi>|^2)}{N} \right\}^{\frac{1}{2}} , \\ \Delta_{\psi^N} F_\ell &<< <F_\ell>_{\psi^N} \quad \text{for} \quad N >> 1 . \end{aligned} \tag{1.18}$$

*If* we could assume the usual interpretation for $<F_\ell>_{\psi^N}$ and $\Delta_{\psi^N} F_\ell$, the above results would be sufficient to "establish" the probability interpretation. We could then say that for a system of N identical particles in the state $\psi$, the average value of the relative frequency operator $F_\ell$ is $|<\psi_\ell, \psi>|^2$. Furthermore, since $\Delta_{\psi^N} F_\ell << <F_\ell>_{\psi^N}$ for large N, the actual observed values of $F_\ell$ would be clustered closely around $|<\psi_\ell, \psi>|^2$.

Of course we cannot assume the usual interpretation for $\langle F_\ell \rangle_{\psi^N}$ and $\Delta_{\psi^N} F_\ell$ if our aim is to establish just that interpretation. If we are to make further progress based on the above results, we must find some way to interpret the quantities involved independently of the interpretation we are trying to prove.

Now there is one well known situation in physics where the mean and standard deviation of a quantum observable are calculated without using the probability interpretation. This situation occurs in statistical mechanics. Consider a macroscopic system in thermal equilibrium with its surroundings and let $X_i$ for $i = 1$ to $n$ be the eigenvalues of $X$ accessible to the system. (In practice $n$ will be a very large but finite number.) Then the average and standard deviation of $X$ are calculated by

$$\begin{aligned} \bar{X} &\equiv \frac{1}{n} \sum_{i=1}^{n} X_i \, , \\ \delta X &\equiv \left\{ \frac{1}{n} \sum_{i=1}^{n} (X_i - \bar{X})^2 \right\}^{\frac{1}{2}} . \end{aligned} \tag{1.19}$$

That is, at least for the overwhelming majority of possible states $\psi$ of the macroscopic system, $\langle X \rangle_\psi$ and $\Delta_\psi X$ are equal (at least approximately) to $\bar{X}$ and $\delta X$.

It is of interest to consider this result in light of the Everett interpretation. If a measurement is made of $X$, then the state of the system and apparatus will split into Everett worlds, one for each of the eigenvalues $X_i$ of $X$. Then $\bar{X}$ and $\delta X^2$ are averages over all the Everett worlds present after the observation. If $\delta X << \bar{X}$, then $X_i$ will be near $\bar{X}$ in the overwhelming majority of these worlds. On the other hand, if $\delta X \sim |\bar{X}|$, then the $X_i$ will take on diverse values in the different worlds.

Based on these results, then, our plan is as follows:

1. Use a macroscopic apparatus to measure the relative frequency operator on a collection of identically prepared systems. Show that after the measurement $X = g\,|\langle \psi_\ell, \psi \rangle|^2$, and $\delta X << |\bar{X}|$, where $X$ is a suitably defined apparatus variable and $g$ is a scale factor.

2. Let an observer read the apparatus (that is, measure X). This interaction will cause the state of the observer and the apparatus to split into Everett worlds. Since $\bar{X}$ and $\delta X^2$ are just averages of the $X_i$ over these worlds, and since $\delta X << |\bar{X}|$ the observed pointer readings, $X_i$, will be near $\bar{X} = g|<\psi_\ell, \psi>|^2$, except in a small minority of the worlds in question.

We begin the implementation of these ideas with a study of some properties of the relative frequency operator.

## 2. *The relative frequency operator*

Using the definitions of $F_\ell$ and $\psi^N$, the quantities $<F_\ell>_{\psi^N}$ and $\Delta_{\psi^N} F_\ell$ can be written

$$<F_\ell>_{\psi^N} = \sum_{m_1 \ldots m_N} f_\ell(m_1, \ldots, m_N) |<\psi_{m_1}, \psi>|^2 \ldots |<\psi_{m_N}, \psi>|^2 ,$$

and (2.1)

$$(\Delta_{\psi^N} F_\ell)^2 = \sum_{m_1 \ldots m_N} (f_\ell - <F_\ell>_{\psi^N})^2 |<\psi_{m_1}, \psi>|^2 \ldots |<\psi_{m_N}, \psi>|^2 .$$

Note that the averages are with respect to the measure of (1.13). This is just Everett's measure; it is through (and only through) $<F_\ell>_{\psi^N}$ and $\Delta_{\psi^N} F_\ell$ that this measure enters the present theory. Using (1.14) we can write

$$<F_\ell>_{\psi^N} = \sum_R \frac{R}{N} \binom{N}{R} P^R Q^{N-R} ,$$

and (2.2)

$$(\Delta_{\psi^N} F_\ell)^2 = \sum_R \left(\frac{R}{N} - <F_\ell>_{\psi^N}\right)^2 \binom{N}{R} P^R Q^{N-R} ,$$

that is, $<F_\ell>_{\psi^N}$ and $\Delta_{\psi^N} F_\ell$ are $\frac{1}{N}$ times the mean and standard deviation of a Bernoulli distribution. Thus $<F_\ell>_{\psi^N} = P$ and $\Delta_{\psi^N} F_\ell = \sqrt{\frac{PQ}{N}}$

Rather than simply invoking this statistical result, it is worthwhile to derive it directly, and in the notation of quantum mechanics. This is most easily done with the moment generating function.

$$<e^{F_\ell u}>_{\psi^N}\,. \tag{2.3}$$

Using this function, we have

$$<F_\ell>_{\psi^N} = \frac{d}{du}<e^{F_\ell u}>_{\psi^N}\Big|_{u=0}\,,$$

and (2.4)

$$<F_\ell^2>_{\psi^N} = \frac{d^2}{du^2}<e^{F_\ell u}>_{\psi^N}\Big|_{u=0}\,.$$

We can then calculate $\Delta_{\psi^N} F_\ell$ from

$$\begin{aligned}\Delta_{\psi^N} F_\ell &= \left[<(F_\ell - <F_\ell>_{\psi^N})^2>_{\psi^N}\right]^{\frac{1}{2}} \\ &= \left[<F_\ell^2 - 2<F_\ell>_{\psi^N} F_\ell + <F_\ell>^2_{\psi^N}>\right]^{\frac{1}{2}} \\ &= \left[<F_\ell^2>_{\psi^N} - <F_\ell>^2_{\psi^N}\right]^{\frac{1}{2}}\,.\end{aligned} \tag{2.5}$$

The advantage of $e^{F_\ell u}$ is that the evaluation of $<e^{F_\ell u}>_{\psi^N}$ is straightforward for the state of interest. Remembering that $\psi^N = \psi \ldots \psi$ (N terms) $= \sum_{m_1 \ldots m_N} <\psi_{m_1}, \psi> \ldots <\psi_{m_N}, \psi>\, \psi_{m_1} \ldots \psi_{m_N}$, we have

$$\begin{aligned}<e^{F_\ell u}>_{\psi^N} &= \sum_{m_1 \ldots m_N} e^{\frac{u}{N}\sum_k \delta_{\ell m_k}} |<\psi_{m_1}, \psi>|^2 \ldots |<\psi_{m_N}, \psi>|^2 \\ &= \sum_{m_1 \ldots m_N} \prod_{k=1}^{N} e^{\frac{u}{N}\delta_{\ell m_k}} |<\psi_{m_k}, \psi>|^2 \\ &= \prod_{k=1}^{N} \sum_{m_k} e^{\frac{u}{N}\delta_{\ell m_k}} |<\psi_{m_k}, \psi>|^2 \\ &= \left(\sum_{m_k} e^{\frac{u}{N}\delta_{\ell m_k}} |<\psi_{m_k}, \psi>|^2\right)^N \\ &= \left(1 - |<\psi_\ell, \psi>|^2 + |<\psi_\ell, \psi>|^2\, e^{\frac{u}{N}}\right)^N\,,\end{aligned} \tag{2.6}$$

where we have used

$$\sum_{m_k \neq \ell} |\langle\psi_{m_k}, \psi\rangle|^2 = 1 - |\langle\psi_\ell, \psi\rangle|^2 \tag{2.7}$$

in the last step.

By substituting the result of (2.6) into (2.4) and (2.5), we get

$$\langle F_\ell\rangle_{\psi^N} = |\langle\psi_\ell, \psi\rangle|^2 \ ,$$

$$\langle F_\ell^2\rangle_{\psi^N} = \frac{N-1}{N} |\langle\psi_\ell, \psi\rangle|^4 + \frac{1}{N} |\langle\psi_\ell, \psi\rangle|^2 \ , \tag{2.8}$$

$$\Delta_{\psi^N} F_\ell = \left[\frac{|\langle\psi_\ell, \psi\rangle|^2 (1 - |\langle\psi_\ell, \psi\rangle|^2)}{N}\right]^{\frac{1}{2}} \sim \frac{1}{\sqrt{N}} \ .$$

Thus $\langle F_\ell\rangle_{\psi^N} = |\langle\psi_\ell, \psi\rangle|^2$ and $\Delta_{\psi^N} F_\ell \ll \langle F_\ell\rangle_{\psi^N}$ for large N. *If* $\Delta_{\psi^N} F_\ell \ll \langle F_\ell\rangle_{\psi^N}$ implies that a measurement of $F_\ell$ will yield a value near $|\langle\psi_\ell, \psi\rangle|^2$, *then* we have established the probability interpretation. This result hinges, however, on establishing the premise of the last statement. To this end we turn to the analysis of the macroscopic apparata that can be used to measure relative frequency.

## 3. *Measurements with a macroscopic apparatus*

The properties of $\langle F_\ell\rangle_{\psi^N}$ and $\Delta_{\psi^N} F_\ell$ proved in the previous section are of no immediate use to us since $F_\ell$ is a microscopic variable, whereas what we need is a macroscopic variable with the same properties. An apparatus serves to transmit the important properties of $F_\ell$ to a macroscopic apparatus variable, X.

Let us consider a measurement of a system variable, (with eigenvalues m and eigenstates $\psi_m$). The initial state of the apparatus is constrained to lie in a certain space $\mathcal{H}_0$. This space is of very large dimension, but represents the greatest practical restriction that can be placed on the initial apparatus state. (To speak of a single possible initial apparatus state is pure fantasy; we can no more prepare a macroscopic system in some specific quantum state than we could fix the position and momentum of every particle in a macroscopic classical system.)

After the measurement has taken place, the state of the system and the apparatus lies in the space

$$\underset{m}{\oplus} \mathcal{H}_m{}^f \tag{3.1}$$

with $\mathcal{H}_m{}^f$ defined by

$$\mathcal{H}_m{}^f = U(\psi_m \otimes \mathcal{H}_0) \tag{3.2}$$

where $U$ is the unitary operator describing the measurement interaction. The spaces $\mathcal{H}_m{}^f$ are distinct, and even orthogonal, but this is because of their dependence on the initial system state $\psi_m$. Whether or not the measurement has produced any change in the large scale properties of the apparatus is another matter altogether.

For a moment let us assume the ideal case: there exists a macroscopic variable $X$ which perfectly distinguishes the spaces $\mathcal{H}_m{}^f$. Then $X\psi_m{}^f = x_m\psi_m{}^f$ for $\psi_m{}^f$ a member of $\mathcal{H}_m{}^f$, and $x_{m_1} \neq x_{m_2}$ for $m_1 \neq m_2$. Thus $x_m$ is a one to one function of $m$. For simplicity, we can take this function as a direct proportionality, $x_m = gm$, where $g$ is a scale factor. Under these assumptions we compute $\langle X \rangle_{\psi^f}$ and $\Delta_{\psi^f} X$ to be

$$\begin{aligned}\langle X \rangle_{\psi^f} &= \sum_m |\langle \psi_m, \psi \rangle|^2 x_m \\ &= g \sum_m |\langle \psi_m, \psi \rangle|^2 m \\ &= g \langle M \rangle_\psi ,\end{aligned} \tag{3.3}$$

and

$$\begin{aligned}\Delta_{\psi^f} X &= \sum_m |\langle \psi_m, \psi \rangle|^2 (x_m - \langle X \rangle_{\psi^f})^2 \\ &= g^2 \sum_m |\langle \psi_m, \psi \rangle|^2 (m - \langle M \rangle_\psi)^2 \\ &= g^2 (\Delta_\psi M)^2, \quad \text{or} \quad \Delta_{\psi^f} X = g \Delta_\psi M .\end{aligned}$$

These are the desired results, but we must be prepared to justify them under far more adverse conditions than those assumed above. For one thing, the assumption that the eigenvalues of X faithfully map those of M is unrealistic; every real apparatus has a finite range and a limited accuracy. A more reasonable assumption is the following:

$$\begin{aligned} x_m &= x_a \,, \quad gm < x_a \,; \\ x_m &= x_b \,, \quad gm > x_b \,; \\ x_{m_1} &= x_{m_2} \,, \quad |gm_1 - gm_2| < \delta \,; \end{aligned} \tag{3.4}$$

where $x_a$ and $x_b$ specify the range of the measuring instrument and $\delta$ specifies its accuracy.

Another difficulty is that the exact relation between the eigenvalues of X and M will depend on the particular initial state of the apparatus; different members of $\mathcal{H}_0$ will give different results. Indeed, for some initial states the apparatus will go completely awry and there will be no correlation whatever. We will assume, however, that for the overwhelming majority of initial states the apparatus will perform as desired, and a mapping similar to that of (3.4) will be obtained. (Otherwise we must conclude that the restrictions defining $\mathcal{H}_0$ were ill chosen.)

Now let us return to (3.3), which is important for later work. We are primarily interested in the case where $\Delta_\psi M$ is small compared to the range of the instrument; the probability interpretation, once proved, will take care of the other cases. Thus we assume that $g\Delta_\psi M << |x_b - x_a|$, and that $g<M>_\psi$ lies well within this range. Then by Chebyshev's theorem (discussed in the next section), $|<\psi_m, \psi>|^2$ will be small for values of m such that gm lies outside the range of the instrument. Thus those parts of the sums in (3.3) for which gm lies outside the range $x_a$ to $x_b$ can be neglected. Inside this range, however, $x_m$ is approximately equal to gm, so we have

$$\langle X\rangle_{\psi^f} \approx g\langle M\rangle_\psi \,,$$

and (3.5)

$$\Delta_{\psi^f}X \approx g\Delta_\psi M \,,$$

where the approximation is reasonably good for a properly functioning apparatus.

We will not observe X immediately after the interaction, but only after the apparatus has reached thermal equilibrium (as is required for the developments of the next section). Thus the state of the apparatus immediately after the measurement interaction will undergo time development before being observed. We must require, then, that $\langle X\rangle_{\psi^f}$ and $\Delta_{\psi^f}X$ be preserved by this time development. Indeed, if $\langle X\rangle_{\psi^f}$ changes then X has simply been badly chosen; it cannot retain its "reading" until an observation can be made of it. In the same way we must assume that $\Delta_{\psi^f}X$ retains at least its original order of magnitude. Otherwise the apparatus is subject to spontaneous splitting (due to coupling to some internal quantum process) and again is unsuitable for measurement purposes.

Finally, let us state our results in terms of the operator of actual interest, $F_\ell$. Assume the apparatus measures $F_\ell$ and displays its results with the apparatus variable X, which is defined as in the previous discussion. After the measurement interaction has taken place and the apparatus has come to equilibrium, we have

$$\langle X\rangle_{\psi^f} \approx g|\langle\psi_\ell,\psi\rangle|^2 \,,$$

and (3.6)

$$\Delta_{\psi^f}X \approx g\Delta_{\psi^N}F_\ell$$

for the majority of possible initial states of the apparatus. For N sufficiently large, we have

$$\Delta_{\psi^f}X << |\langle X\rangle_{\psi^f}| \,. \tag{3.7}$$

In Section 5 we will see how (3.7) can be interpreted in terms of the Everett worlds produced by an observation of the apparatus.

## 4. *Some properties of macroscopic systems*

At this point in our argument we need to use some of the basic statistical properties of macroscopic systems. Because of their fundamental importance in the analysis of the probability interpretation, we include here a brief account of the known results in this area. Further discussion and additional references can be found in Reference [5].

Consider a macroscopic system whose state vector, $\psi$, is constrained to lie in a certain space $\mathcal{H}$. If we assume a system confined to a finite region of space, and able to extract at most a finite amount of energy from its surroundings, then $\mathcal{H}$ will be at most finite dimensional. Its dimension, $n$, will of course be a very large number.

Now suppose that at some instant we take a "snapshot" of the system in question. The state of the system seen in this snapshot will be some $\psi$ in $\mathcal{H}$. It is totally beyond our powers to predict this state in detail. We can study it statistically, however, with the aid of the well known ergodic principle: "All the states accessible to a system in thermal equilibrium (all the states in $\mathcal{H}$) are equally likely." This statement of the principle is (to the best of the author's knowledge) somewhat more general than what has been strictly proved. Nevertheless, the principle as stated is widely used in statistical mechanics, and with complete success.

The problem of taking a snapshot of the system at a certain time, then, is statistically equivalent to the one of drawing a state from the space $\mathcal{H}$ at random, with equal chances for each state. To study this random process, we need a measure on the (normalized) states of $\mathcal{H}$ such that all such states are equally likely. This problem, however, has a straightforward geometrical solution.

To begin, choose a basis $\psi_1, \ldots, \psi_n$ in $\mathcal{H}$ and let the state $\psi$ have the expansion

$$\psi = \sum_{j=1}^{n} \langle \psi_j, \psi \rangle \psi_j \ , \tag{4.1}$$

where

$$\sum_{j=1}^{n} |\langle\psi_j, \psi\rangle|^2 = 1 \ . \tag{4.2}$$

If we put

$$\langle\psi_j, \psi\rangle = x_j + iy_j \tag{4.3}$$

then

$$\sum_{j=1}^{n} x_j^{\,2} + \sum_{j=1}^{n} y_j^{\,2} = 1 \ . \tag{4.4}$$

That is, there is a one to one correspondence between points on a unit 2n sphere and the normalized states in $\mathcal{H}$.

Let us define the measure of any set of states S by

$$\mu(S) = \frac{\text{surface area of S on 2n sphere}}{\text{surface area of 2n sphere}} \ . \tag{4.5}$$

Clearly all the $\psi$'s are equally likely under this measure.

If f is any function of $\psi$, we define the mean, variance, and standard deviation of f with respect to $\mu$ as follows:

$$M[f] = \int f(\psi)\, d\mu(\psi) \ ,$$

$$V[f] = M[f^2 - M[f]^2] \ , \tag{4.6}$$

$$S[f] = V[f]^{\frac{1}{2}} \ .$$

Finally, we shall need Chebyshev's inequality. Since

$$M[f^2] = \int f(\psi)^2 d\mu(\psi) \geq \int_{|f(\psi)|>\varepsilon} f(\psi)^2 d\mu(\psi) > \varepsilon^2 \int_{|f(\psi)|>\varepsilon} d\mu(\psi)$$

$$= \varepsilon^2 \mu\{\psi : |f(\psi)| > \varepsilon\} \ ,$$

we have

$$\mu\{\psi : |f(\psi)| > \varepsilon\} < \frac{M[f^2]}{\varepsilon^2} \ . \tag{4.8}$$

In particular,

$$\mu\left\{\psi : \left|\frac{f(\psi)-M[f]}{M[f]}\right| > \varepsilon\right\} < \frac{V[f]}{\varepsilon^2 M[f]^2} . \qquad (4.9)$$

Thus if $V[f] << \varepsilon^2 M[f]^2$ then $f(\psi) \approx M[f]$ except on a set of $\psi$'s of small measure.

(Note that Chebyshev's inequality is quite general and has applications outside the present section. In the last section we applied it to the measure $|<\psi_m, \psi>|^2$ and in the next section we shall apply it to the constant measure $\frac{1}{n}$.)

Let us apply these results to $<X>_\psi$. We have

$$\begin{aligned} M[<X>_\psi] = M[<\psi, X\psi>] &= M\left[\sum_{ij} <\psi, \psi_i> X_{ij} <\psi_j, \psi>\right] \\ &= \sum_{ij} X_{ij} M[<\psi, \psi_i> <\psi_j, \psi>] , \end{aligned} \qquad (4.10)$$

where $X_{ij} = <\psi_i, X\psi_j>$.

If $i \neq j$, then $<\psi, \psi_i> <\psi_j, \psi>$ contains the factor $e^{i(\theta_i - \theta_j)}$ whose average over all $\theta_i, \theta_j$ is zero. Thus

$$M[<\psi, \psi_i> <\psi_j, \psi>] = k\delta_{ij} . \qquad (4.11)$$

Since $\sum_i |<\psi, \psi_i>|^2 = 1$, $1 = kn$, or $k = \frac{1}{n}$.
Hence

$$M[<\psi, \psi_i> <\psi_j, \psi>] = \frac{1}{n}\delta_{ij} , \qquad (4.12)$$

and

$$M[<X>_\psi] = \frac{1}{n} \sum_{i=1}^{n} X_{ii} = \frac{1}{n} \operatorname{tr}(X) . \qquad (4.13)$$

If the $\psi_i$ are eigenstates of X, then this becomes

$$M[<X>_\psi] = \frac{1}{n} \sum_{i=1}^{n} X_i = \overline{X} . \qquad (4.14)$$

Thus the average over all states $\psi$ of the macroscopic system of $<X>_\psi$ equals the average of all the eigenvalues of X.

We now want to know how likely we are to find $<X>_\psi$ near $M[<X>_\psi]$ for a $\psi$ drawn at random from $\mathcal{H}$. To this end, we compute $M[<X>_\psi^2]$ by

$$M[<X>_\psi^2] = \sum_{ijk\ell} X_{ij} X_{k\ell} M[<\psi,\psi_i><\psi,\psi_k><\psi_j,\psi><\psi_\ell,\psi>] . \quad (4.15)$$

Now $M[<\psi,\psi_i><\psi,\psi_k><\psi_j,\psi><\psi_\ell,\psi>]$ will be zero unless ik is a permutation of $j\ell$, since otherwise there will be an uncanceled phase factor whose average over all equally likely phase angles is zero. Since the average is symmetrical under the interchange of i and k or j and $\ell$, we must have

$$M[<\psi,\psi_i><\psi,\psi_k><\psi_j,\psi><\psi_\ell,\psi>] = K(\delta_{ij}\delta_{k\ell} + \delta_{i\ell}\delta_{kj}) . \quad (4.16)$$

Using $\sum_j |<\psi_j,\psi>|^2 \sum_\ell |<\psi_\ell,\psi>|^2 = 1$ gives $1 = Kn(n+1)$ thus

$$\begin{aligned} M[<\psi,\psi_i><\psi,\psi_k><\psi_j,\psi><\psi_\ell,\psi>] &= \frac{\delta_{ij}\delta_{k\ell} + \delta_{i\ell}\delta_{kj}}{n(n+1)} \\ &\approx \frac{1}{n^2}(\delta_{ij}\delta_{k\ell} + \delta_{i\ell}\delta_{kj}) . \end{aligned} \quad (4.17)$$

Then

$$M[<X>_\psi^2] = \frac{\mathrm{tr}(X)^2 + \mathrm{tr}(X^2)}{n(n+1)} \approx \frac{1}{n^2}[\mathrm{tr}(X)^2 + \mathrm{tr}(X^2)] . \quad (4.18)$$

The variance of $<X>_\psi$ is thus given by

$$V[<X>_\psi] = M[<X>_\psi^2] - M[<X>_\psi]^2 \approx \frac{1}{n^2}\mathrm{tr}(X^2) . \quad (4.19)$$

We are now in a position to apply Chebyshev's inequality and get

$$\mu\left\{\psi : \left|\frac{<X>_\psi - M[<X>_\psi]}{M[<X>_\psi]}\right| > \varepsilon\right\} < \frac{\mathrm{tr}(X^2)}{\varepsilon^2\,\mathrm{tr}(X)^2} .$$

Now the trace of a matrix is the sum of n terms and hence of order n. Thus

$$\mathrm{tr}(X) \sim \mathrm{tr}(X^2) \sim n \ , \tag{4.21}$$

and

$$\mu\left\{\psi : \left|\frac{\langle X\rangle_\psi - M[\langle X\rangle_\psi]}{M[\langle X\rangle_\psi]}\right| > \varepsilon\right\} < \frac{1}{n\varepsilon^2} \ . \tag{4.22}$$

Since n is indeed a large number, the set of all $\psi$'s for which $\langle X\rangle_\psi$ differs appreciably from $M[\langle X\rangle_\psi]$ has a very small measure for sufficiently large n.

A similar argument can be given for $\Delta_\psi X$. To start with, let us define $\Delta_\psi X$ by

$$(\Delta_\psi X)^2 = \langle (X - x_0)^2 \rangle_\psi \ , \tag{4.23}$$

where $x_0$ is an arbitrary constant. Then

$$M[(\Delta_\psi X)^2] = M[\langle (X-x_0)^2\rangle_\psi] = \frac{1}{n}\,\mathrm{tr}((X-x_0)^2) \ , \tag{4.24}$$

and, by our previous result

$$(\Delta_\psi X)^2 \approx \frac{1}{n}\,\mathrm{tr}((X-x_0)^2) \tag{4.25}$$

for all $\psi$ except a set of very small measure. If we take $x_0 = \overline{X} = \frac{1}{n}\,\mathrm{tr}(X)$ then

$$\begin{aligned} \langle (X-\overline{X})^2\rangle_\psi &\approx \overline{(X-\overline{X})^2} \\ &\approx (\delta X)^2 \ , \end{aligned} \tag{4.26}$$

except on a set of small measure. However, we also have $\overline{X} \approx \langle X\rangle_\psi$ except on a set of small measure.

Let A be the set of $\psi$'s for which $\overline{X} \not\approx \langle X\rangle_\psi$ and B the set for which $\langle (X-\overline{X})^2\rangle_\psi \not\approx (\delta X)^2$. Then $A \cup B$ is the set on which one or both of the equalities fail. Since

$$\mu(A \cup B) \leq \mu(A) + \mu(B) \ , \tag{4.27}$$

and since $\mu(A)$ and $\mu(B)$ can be made as small as desired, we can make $\mu(A \cup B)$ as small as desired also. Thus, except on $A \cup B$,

$$<(X-\overline{X})^2>_\psi \approx \overline{(X-\overline{X})^2}\ , \tag{4.28}$$

and

$$<X>_\psi \approx \overline{X}\ , \tag{4.29}$$

hence

$$<(X-<X>_\psi)^2>_\psi \approx \overline{(X-\overline{X})^2}\ , \tag{4.30}$$

which is the desired result. Then, except on a set of small measure,

$$<X>_\psi \approx \frac{1}{n}\sum_i X_i = X\ , \tag{4.31}$$

and

$$\Delta_\psi X \approx \left\{\frac{1}{n}\sum_i (X_i-\overline{X})^2\right\}^{\frac{1}{2}} \approx \overline{\{(X-\overline{X})^2\}}^{\frac{1}{2}} \approx \delta X\ . \tag{4.32}$$

5. *Summary and Conclusions*

Now let us describe a complete measurement of the relative frequency of the occurrence of the value $\ell$ for an observable M. We define the relative frequency operator, $F_\ell$, on a collection of N systems all in the same state $\psi$. The state of the collection is $\psi^N = \psi\ldots\psi$ (N terms). The operator $F_\ell$ satisfies

$$<F_\ell>_{\psi^N} = |<\psi_\ell,\psi>|^2\ ,$$

$$\Delta_{\psi^N} F_\ell = \left\{\frac{|<\psi_\ell,\psi>|^2\,(1-|<\psi_\ell,\psi>|^2)}{N}\right\}^{\frac{1}{2}} \sim \frac{1}{\sqrt{N}}\ ,$$

and $\Delta_{\psi^N} F_\ell << <F_\ell>_{\psi^N}$ for large N.

We measure $F_\ell$ with an apparatus having an apparatus variable X. After the measurement interaction is complete, and the apparatus has returned to thermal equilibrium,

$$\langle X\rangle_{\psi_0} \approx g|\langle\psi_\ell,\psi\rangle|^2 \; , \tag{5.2}$$
$$\Delta_{\psi_0} X \approx g\Delta_{\psi^N} F_\ell \; ,$$

and

$$\Delta_{\psi_0} X << \langle X\rangle_\psi \; ,$$

where $\psi_0$ is the final state of the system and apparatus and $g$ is a scale factor.

Now let an observer measure $X$. The final state of the apparatus and the observer is

$$\psi^f = \sum_i \langle\psi_i,\psi_0\rangle\psi_i\phi[X_i] \; , \tag{5.3}$$

where $\psi_i$ is an eigenstate of $X$ with eigenvalue $X_i$, and $\phi[X_i]$ is the state of the observer when it has observed $X_i$. (To be strictly correct, we should consider spaces of observer states, rather than single states, but these spaces play no role in the discussion at this point.)

Thus the final state of the apparatus and observer describes a set of Everett worlds, with one eigenvalue $X_i$ being observed in each world. According to our previous work, however,

$$\overline{X} = \frac{1}{n}\sum_i X_i \approx \langle X\rangle_{\psi^f} \approx g|\langle\psi_\ell,\psi\rangle|^2 \; , \tag{5.4}$$

$$\delta X = \left\{\frac{1}{n}\sum_i (X_i-\overline{X})^2\right\}^{\frac{1}{2}} \approx g\Delta_{\psi^N} F_\ell \sim \frac{g}{\sqrt{n}} \; ,$$

and $\delta X << |\overline{X}|$ for the overwhelming majority of the initial apparatus states. Thus (by Chebyshev's inequality) the overwhelming majority of the $X_i$ are near $\overline{X}$, and values near this are observed in the majority of the corresponding Everett worlds. More specifically, if $n\{P\}$ is the number of Everett worlds in which the condition $P$ holds, and $n$ the total number of such worlds, then

$$\frac{n\left\{\left|\frac{X_i - \overline{X}}{\overline{X}}\right| > \varepsilon\right\}}{n} < \frac{(\delta X)^2}{\varepsilon^2\,\overline{X}^2}\,. \tag{5.5}$$

Since $\delta X << \overline{X}$ then $n\left\{\left|\frac{X_i - \overline{X}}{\overline{X}}\right| > \varepsilon\right\} << n$ for reasonably small values of $\varepsilon$.

We thus conclude that values of relative frequency near $|<\psi_\ell, \psi>|^2$ will be found in the majority of Everett worlds of the apparatus and observer. If we assume our own world to be a "typical" one, then we may expect a human or mechanical observer to perceive relative frequencies in accordance with the Born interpretation. Why we should be able to assume our own world to be typical is, of course, itself an interesting question, but one that is beyond the scope of the present paper.

## REFERENCES


[1] H. EVERETT, Rev. Mod. Phys. 29, 454 (1957).

[2] H. EVERETT, "The Theory of the Universal Wave Function" in *The Many-Worlds Interpretation of Quantum Mechanics*, B. S. DeWitt and N. Graham, eds. (Princeton University Press, Princeton, 1973).

[3] N. GRAHAM, *The Everett Interpretation of Quantum Mechanics* (Ph.D. Thesis, The University of North Carolina, 1970).

[4] J. HARTLE, Am. J. Phys. 36, 704 (1968).

[5] G. PROSPERI and A. SCOTTI, Nuovo Cimento, 17, 267 (1960).

[6] A. DANERI, A. LOINGER, and G. PROSPERI, Nucl. Phys. 33, 297 (1962).

[7] A. DANERI, A. LOINGER, and G. PROSPERI, Nuovo Cimento, Ser. 10, 44B, 119 (1966).

[8] A. GLEASON, J. Math. and Mech. 6, 885 (1957).

**Library of Congress Cataloging in Publication Data**

DeWitt, Bryce Seligman, 1923-        comp.
The many-worlds interpretation of quantum mechanics.

(Princeton series in physics)
Includes bibliographical references.
CONTENTS: Everett, H. The theory of the universal wave function.—Everett, H. "Relative state" formulation of quantum mechanics.—Wheeler, J. A. Assessment of Everett's "Relative state" formulation of quantum theory. [etc.]
1. Quantum theory. I. Everett, Hugh. II. Graham, Neill, 1941- joint comp. III. Title.
QC174.12.D48  530.1'2  72-12116
ISBN 0-691-08126-3

Library of Congress Cataloging in Publication Data

ISBN 0-691-08131-X (pbk.)

Milton Keynes UK
Ingram Content Group UK Ltd.
UKHW010840140924
448309UK00008B/328

9 780691 618951